Technical English 1

Course Book

David Bonamy

PEARSON
Longman

Contents

Unit	Section	Topics	Language	Vocabulary
Unit 1 Check-up	1.1 **Basics** p.4	Meeting and greeting people Using forms Following instructions	Verb *be* *I'm Danielle. I'm a technician.* Imperatives	Basic vocabulary: *say, write* … Tools, fixings, electrical parts, occupations
	1.2 **Letters and numbers** p.6	Exchanging information Using forms Units of measurement	*How do you spell …?*	Personal details Cardinal numbers Alphabet Abbreviations of units: *kg, m*
	1.3 **Dates and times** p.8	Using numbers Talking about travel timetables Making appointments	*LH three oh six. Monday the twenty-eighth of December.*	Ordinal numbers Dates and times Decimals
Unit 2 Parts (1)	2.1 **Naming** p.10	Identifying things	*What's that? I think it's a car.* *What's this called?* *this, that, these, those*	Parts: *wheel, axle, plate* … Fixings: *nuts, bolts, nails* … Vehicles: *car, bike, plane* …
	2.2 **Assembling** p.12	Using checklists Saying what you need for a job Using an instruction manual	*I need some bolts. What size?* Imperative + object + location: *Put the wheel on the axle.*	Verbs: *loosen, tighten, push* … Linear: *mm, mil, millimetre*
	2.3 **Ordering** p.14	Using voicemail Ordering by phone Introducing yourself and others	*How many do you need?* *What's your name? Please spell that.*	Numbers: *double 5, zero* Sizes: *small, medium, large* Colours: *red, blue, black* …
Review Unit A p.16				
Unit 3 Parts (2)	3.1 **Tools** p.20	Describing components Using a product review	Present simple of *have* *My multi-tool has blades and a spanner.*	Tools: *spanner, (a pair of) pliers* Parts of tools: *shaft, blade, head* …
	3.2 **Functions** p.22	Saying what things do Describing a product Talking about people's jobs	Present simple *What does this handle do?* *Where do you work?*	Verbs: *measure, grip, cut, open* … Everyday tools: *torch, alarm* … Occupations: *operator, technician* …
	3.3 **Locations** p.24	Saying where things are	Adverbials and prepositions of location *Where is it? It's at the top.*	Location: *top, bottom, middle* … Computer and electronic equipment
Unit 4 Movement	4.1 **Directions** p.26	Describing direction of movement	Adverbials of direction *can, can't, cannot* *Can a helicopter fly backwards? Yes, it can.*	Direction: *up, down, forwards* … Adverbs: *straight, vertically* Angles: *degrees*
	4.2 **Instructions** p.28	Using an instruction manual	Imperative + present simple *Push the joystick upwards and the plane accelerates.*	Movement: *ascend, descend* … Controls: *joystick, slider* … Speed: *km/h, m/s*
	4.3 **Actions** p.30	Using an instruction manual Giving and following instructions Explaining what happens	*When* clause *When you pull the lever backwards, the truck reverses.*	Movement: *drive, reverse* …
Review Unit B p.32				
Unit 5 Flow	5.1 **Heating system** p.36	Explaining how fluids move around a system Using a flow chart	Present simple *The water flows through the pipe into the tank.* Prepositions of movement	Parts of a fluid system: *inlet* … Prepositions: *into, out of, to* … Verbs: *enter, flow, sink* …
	5.2 **Electrical circuit** p.38	Explaining how an electrical circuit works	Zero conditional *If the battery is empty, the current doesn't flow.*	Circuit: *battery, conductor* … Electrical units: *ampere, watt*
	5.3 **Cooling system** p.40	Explaining how cooling systems work Describing everyday routine	Reference words: *here, it, this* Present simple in routines	Cooling system parts: *engine, fan* … Temperature: *degrees Celsius*
Unit 6 Materials	6.1 **Materials testing** p.42	Giving a demonstration Explaining what you're doing	Present continuous *I'm stretching the rope.*	Verbs: *bend, cut, compress* … Spelling: *strike/striking* …
	6.2 **Properties** p.44	Describing the properties of materials	*What's it made of?* *You can't bend it. = It's rigid.*	Materials: *aluminium, graphite* … Properties: *hard, rigid, tough* …
	6.3 **Buying** p.46	Using a customer call form Buying and selling by phone Checking Starting a phone call	*What's your email address?* *Could you spell/repeat that?* *How many would you like?*	Email/Web addresses: *at, dot* … Prices: *euro, dollar*
Review Unit C p.48				

Grammar summary p.100   **Reference section p.106**

Unit	Section	Topics	Language	Vocabulary
Unit 7 Specifications	7.1 **Dimensions** p.52	Specifying dimensions Using a specifications chart	*How long is it? It's 9 mm long.* *The length of the road is 120 km.*	Bridge parts: *deck, pier, pylon* Adjectives/nouns: *long/length, high/height* Linear and weight: *mm, m, kg …*
	7.2 **Quantities** p.54	Specifying materials Buying materials for a job Using a materials checklist	Countable and uncountable nouns *I'd like some paint, please.*	Substances: *glue, cement, oil …* Containers: *tube, tin, bag …* Area and volume: m^2, m^3, *litre …*
	7.3 **Future projects** p.56	Describing plans for the future Using a Gantt chart	*will, won't* Time expressions: *in 2015, at the end of 2015*	Verbs: *attach, complete, connect …*
Unit 8 Reporting	8.1 **Recent incidents** p.58	Taking an emergency call Explaining what has happened Checking on progress	Present perfect *I've checked the brakes.* *Have you checked the tyres?*	Car repair: *brakes, exhaust pipe …* Building site: *beam, bucket, digger …*
	8.2 **Damage and loss** p.60	Reporting damage Dealing with a customer	Past participles as adjectives: *It's broken.* *They're dented.* *There are some scratches on the screen.* *There's no user manual.*	Electrical: *antenna, plug …* Damage: *bent, broken, dented …* Loss: *missing …*
	8.3 **Past events** p.62	Discussing past events Phoning a repair shop	Past simple *They launched it in 2006.* Time expressions: *in 2008, on 5th October, fifty years ago …*	Time: *today, yesterday, a week ago …* Revision of dates and years *more than, less than*
Review Unit D p.64				
Unit 9 Troubleshooting	9.1 **Operation** p.68	Explaining how things work Explaining what things do	Revision of present simple *The handlebar steers the airboard.*	Verbs: *control, drive, press …* Parts: *body, lever …* Connections: *attached to, mounted on …*
	9.2 **Hotline** p.70	Listening to an automated phone message Using a service hotline Taking a customer through a problem and solution	*Is the computer connnected to the adapter?* Short answers: *Yes, I have. No, it doesn't.* *Yes, it is.*	Electronics and computing: *RF/SCART socket, router, modem…* Connections: *connected to*
	9.3 **User guide** p.72	Using a flow chart Using a troubleshooting guide	Zero conditional + imperative *If it doesn't start, check the cable.*	Electronics: *LED, loose (cable) …* Computing: *disk drive, printer …* Car repair: *flat (battery) …*
Unit 10 Safety	10.1 **Rules and warnings** p.74	Following safety rules Giving and following warnings Using safety signs	*could, might, must* *Always… Don't… You mustn't…* *You might trap your hand.*	Safety gear: *hard hat, gloves …* Hazards: *poison, danger …* Accidents: *hurt, injure, trap …* Shapes: *circular, round …*
	10.2 **Safety hazards** p.76	Giving and following warnings Noticing safety hazards Reporting safety hazards	Past tense of *be* *The fire exit was locked.* *There were no fire extinguishers.*	Hazard nouns: *gap, bare wire …* Hazard adjectives: *coiled, damaged, locked …* Safety: *fire exit, safety cone …*
	10.3 **Investigations** p.78	Investigating an accident Reporting an accident Giving, accepting and turning down an invitation	Questions in the past simple *Where? When? How high? What? How far? How many?*	Nouns on a form: *position, altitude, distance …*
Review Unit E p.80				
Unit 11 Cause and effect	11.1 **Pistons and valves** p.84	Expressing causation, permission and prevention Explaining how a four-stage cycle works	Verb constructions *cause, allow* + *to* infinitive *make, let* + bare infinitive *stop, prevent* + *from* + gerund	Hydraulics: *chamber, inlet, outlet …*
	11.2 **Switches and relays** p.86	Explaining how a relay circuit works Giving an oral presentation	Further practice of verb patterns in 11.1	Electrical: *battery, buzzer, earth …*
	11.3 **Rotors and turbines** p.88	Explaining how a wind turbine works Giving an oral presentation Making suggestions	Further practice of verb patterns in 11.1 Reference words: *it, one*	Turbines: *blade, brake, gear …* Verbs: *drive, rotate, send …*
Unit 12 Checking and confirming	12.1 **Data** p.90	Describing specifications Expressing approximation Checking that data is correct	Revision of question forms *Is that correct? No, that's wrong.*	Approximation: *about, over, at least …* Nouns: *mass, rotation*
	12.2 **Instructions** p.92	Following spoken instructions Confirming actions Describing results of actions	Revision of imperative with present continuous	Revision of controls, vehicles, direction adverbs, verbs of movement
	12.3 **Progress** p.94	Describing maintenance work Checking progress with a Gantt chart	Revision of present perfect, past simple, present continuous, and *will*	Maintenance and repair: *check, inspect, assemble …*
Review Unit F p.96				

Extra material p.112

Audio script p.119

1 Check-up

1 Basics

Start here 1 🔊 02 Listen and complete the dialogues with the words in the box.

| am | are | I'm | is | name's |

1 ● *Hello. I (1) <u>am</u> Hans Beck.*
  ○ *Hi. My name (2) _____ Pedro Lopez.*
  ● *Pleased to meet you.*
2 ● *Excuse me. (3) _____ you Mr Rossi?*
  ○ *Yes, I am.*
  ● *Pleased to meet you, Mr Rossi. (4) _____ Danielle Martin.*
  ○ *Nice to meet you, Danielle.*
3 ● *Hi. My (5) _____ Jamal.*
  ○ *Hello, Jamal. (6) _____ Borys.*
  ● *Good to meet you, Borys. (7) _____ you from Russia?*
  ○ *No, (8) _____ from Poland.*

I am → I'm
My name is → My name's
What is → What's

2 Work in pairs. Practise the dialogue in 1 with your partner. Talk about yourself.

Writing 3 Complete the form about yourself. Use block capitals.

Name	Country	Occupation
___	___	___

Speaking 4 Work in pairs. Ask and answer questions.

A: *Hello. What's your name?*   B: *I'm Kato.*
A: *Where are you from?*   B: *I'm from Japan.*
A: *What do you do?*   B: *I'm a builder/an electrician/a student.*

What do you do? = What's your job/occupation?

Listening 5 ▶️03 Play this game. Listen. Only follow the instructions if the speaker says *Please*.

Vocabulary 6 Match the opposites.

| pick up | raise | read | say | stand | start |

| listen | lower | put down | sit | stop | write |

Example: stand ≠ sit

7 Try this quiz. Choose the correct answer.

  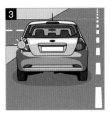   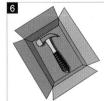

1 The TV is   a) on.   b) off.
2 The doors are   a) closed.   b) open.
3 Turn   a) left.   b) right.
4 Go   a) in.   b) out.
5 Drive   a) up.   b) down.
6 The hammer is   a) in the box.   b) on the box.   c) under the box.

8 Match the pictures with the words in the box.

| adapter   antenna   bolts   cable   chisel   nuts   plug |
| saw   screws   screwdriver   spanner   washers |

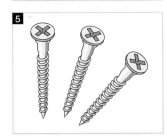 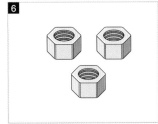  

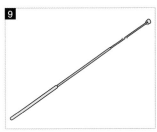   

2 Letters and numbers

Start here 1 ▶ 04 Listen and correct the four mistakes in the business card.

Listening 2 ▶ 05 Listen and complete the forms.

1

Surname	__ __ A __ L __
Company	__ __ __ __
Email address	__ q __ __ @ __ __ __ .com

2

Emergency service	FIRE
Address	__ __ E __ S __ Street
Postcode	__ __ 4 __ N __
Surname	__ A T __ E __ S

3 Problems with your product?
Phone CUSTOMER SERVICE HELPLINE

Full name	PIETER __ __ R __ U __
Postcode	2 __ __ __ __
House number	__ __
Model number	__ __ 8 __ __ __

Speaking 3 Dictate and spell out details from your business card to your partner.

4 Put all the letters of the alphabet into the correct column.

three	eight	five	ten	two	EXCEPTIONS
B _ _ _ _ _ _ _	A _ _ _	I _	F _ _ _ _ _ _	Q _ _	_ _

5 Work in groups. Have a spelling competition.

Team A: Make a list of ten countries. Check the spelling. Then ask Team B to spell them correctly.
Team B: Make a list of ten capital cities. Check the spelling. Then ask Team A to spell them correctly.

Example: How do you spell EGYPT? How do you spell TOKYO?

Listening 6 🔊 06 Listen and match the pictures with the announcements.

7 Listen again and complete the sentences with numbers and letters.

1. Counter number _____, please.
2. This is Radio _____ on _____ FM.
3. Please pay _____ pounds and _____ pence.
4. The _____ train to Oxford will depart from platform number ___.
5. Flight number _____ is boarding now. Please go to gate number ___.
6. To donate money to Live Aid, ring this number now: _____.
7. Begin countdown now: _____ …

Speaking 8 Play *FIZZ BUZZ*.

- Count from 1 to 100 round the class.
- Use *Fizz* for a number you can divide by 3. *Example: 3, 6, 9, 12, …*
- Use *Buzz* for a number you can divide by 5. *Example: 5, 10, 15, 20, …*
- Use *Fizz Buzz* for a number you can divide by both 5 and 3. *Example: 15, 30, …*
- If you make a mistake, you are OUT of the game.

Start like this: 1, 2, Fizz, 4, Buzz, Fizz, 7, 8, Fizz, Buzz, 11, Fizz, 13, 14 …

Vocabulary 9 What do the following mean?

km + g in kW kg L V
A ° rpm C km/h
m £ – ft € W gal

amps = watts / volts

Example: km = kilometre

Listening 10 🔊 07 Listen and write the numbers in the correct space.

1	_____ °C	5	_____ °	9	_____ W
2	_____ A	6	_____ km/h	10	_____ V
3	_____ km	7	_____ rpm	11	_____ €
4	_____ m	8	_____ kg	12	_____ L

Check-up 1

3 Dates and times

Start here

1 🔊 08 Listen to the sports results. Add the positions (2ⁿᵈ, 3ʳᵈ and 5ᵗʰ) and complete the times in the blanks in the chart.

Athens Olympics 2004 Official Results Men's Finals: 1500 metres			
Position	Name	Country	Time
(1)	Silva	Portugal	3:34.68
4ᵗʰ	Kiptanui	Kenya	(2) 3:___.___
1ˢᵗ	El Guerrouj	Morocco	(3) 3:___.___
(4)	Lagat	Kenya	3:34.30
6ᵗʰ	East	Britain	(5) 3:___.___
(6)	Heshko	Ukraine	3:35.82

Speaking

2 Put the ordinal numbers *1ˢᵗ* to *31ˢᵗ* into the chart. Read them out to your teacher.

-st	-nd	-rd	-th
1ˢᵗ,	2ⁿᵈ,	3ʳᵈ,	4ᵗʰ,

3 Say the names of the months of the year.

4 Say the days of the week. Start with **Monday**.

5 Read out these airport codes.

FRA = Frankfurt	WAW = Warsaw	DXB = Dubai	CAI = Cairo
CDG = Paris	MAD = Madrid	FCO = Rome	NRT = Tokyo
LHR = London	BAH = Bahrain	JNB = Johannesburg	LOS = Lagos

6 Give the days of the flights.

on Mondays = on Monday every week

Flight number	From	To	Depart	Arrive	Days
1 LH 306	FRA	WAW			1 4
2 AF 835	CDG	MAD			2 4 6
3 EK 971	LHR	BAH			1 2 4 5
4 MS 740	DXB	CAI			1 3 5 7
5 AZ 7788	FCO	NRT			2 3 5 6
6 SA 104	JNB	LOS			1 4 7
1 = Monday 2 = Tuesday 3 = Wednesday 4 = Thursday 5 = Friday 6 = Saturday 7 = Sunday					

Example: 1 LH 306 departs from Frankfurt on Mondays and Thursdays.

1 Check-up

Listening   **7**   ▶ 09   Listen and write down the dates. Use *dd/mm/yy*.

Speaking   **8**   Write down some dates important to you. Then dictate them to your partner.

You dictate: *The twenty-eighth of December two thousand and ten.*
Your partner writes: *2010-12-28*.

> 28th December 2010
> - in Europe: *28/12/10* (dd/mm/yy)
> - in the USA: *12/28/10* (mm/dd/yy)
> - in Japan: *10/12/28* (yy/mm/dd)
> - ISO 8601: 2010-12-28 (yyyy-mm-dd)

9   Complete the table. Read out your answers.

write: *0*; say: *oh* or *zero*.

24-hour clock	12-hour clock
07.50	(1) *7.50 am*
17.30	5.30 pm
14.40	(4)
(6)	1.35 pm
05.55	(8)

24-hour clock	12-hour clock
(2)	6.30 am
15.15	(3)
(5)	4.45 pm
20.25	(7)
(9)	9.10 pm

10   Read out these times.

First, use the 24-hour clock. Then use the 12-hour clock.

1) 05.15   2) 08.50   3) 11.14   4) 13.40   5) 15.18   6) 17.30

Listening   **11**   ▶ 10   Listen and add the times to the timetable in 6. Use the 24-hour clock.

12   ▶ 11   Listen and write the correct number next to each watch.

13   Read out the times and dates on the watches in 12. Use the 12-hour clock.

Social English   **14**   Practise this conversation.
Use different days and times.

A: *When's the party?*
B: *It's on Friday.*
A: *Is that Friday the 24th?*
B: *Yes, that's right.*
A: *What time?*
B: *7.30.*
A: *OK. See you then. Bye.*
B: *See you. Bye.*

Invite John to party –
Friday 24th, 7.30

2 Parts (1)

1 Naming

Start here 1 🎧 12 Listen and complete the table.

Skateboard record	Distance	Date (dd/mm/yy)
1 High jump	_____ metres	____/____/____
2 Long jump	_____ metres	____/____/____

Vocabulary 2 Work in pairs. Label the diagram with the words in the box.

axle   deck   nose   plate   tail   truck   wheel

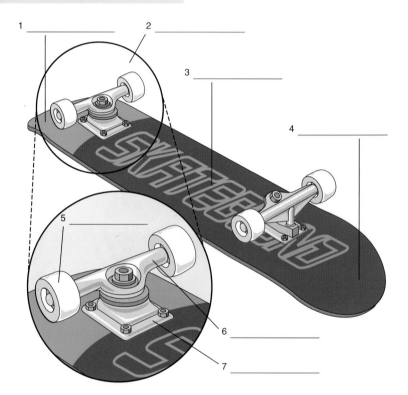

Listening 3 🎧 13 Listen and check your answers to 2.

4 🎧 14 Listen and complete the dialogue.

● *What's this _____?*
○ *It's _____ a deck.*
● *What's _____ called in English?*
○ *It's called _____ truck.*

Speaking 5 Work in pairs. Ask and answer questions about all the parts on the diagram.

A: *What's this called?* (or *What's this called in English?*)
B: *It's called a deck.*

Language *What's this called?* Use this when you don't know the English word.
What's this? Use this when you don't know what it is, even in your own language.

What	's / is	this / that	called	?	It	's / is	called	a	deck.
								an	axle.
What	are	these / those			They	're / are		decks.	
								axles.	

6 Complete the dialogues with the words in the box.

It's that these They're this those

1 ● *What's _____ called in English?*
   ○ _____ called a screw.
2 ● *What's _____ called?*
   ○ _____ called a motorbike.
3 ● *What are _____ called in English?*
   ○ _____ called bolts.
4 ● *What are _____ called?*
   ○ _____ called antennas.

Vocabulary **7** ▶ 15 Listen and repeat.

nails ... bolts ... nuts ... spanner ... washers ... staples ... screws ... screwdriver

8 Match the words from 7 with the pictures.

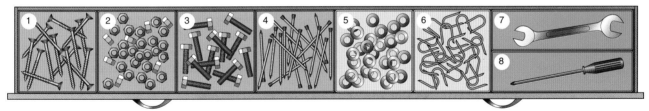

Speaking **9** Work in pairs. Ask and answer questions about the tools and fixings.

A: *What are these called?*

10 Point to things in the class or outside. Ask and answer questions.

What's this/that called? What are these/those called?

11 Work in small groups. What are these?

Clue: they're all vehicles on land, sea, in air and space.

A: *What's this?*

Answers on page 113.

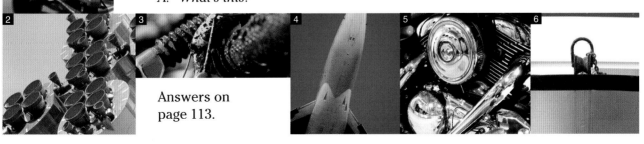

2 Assembling

Start here 1 Work in pairs. You want to assemble a skateboard. What do you need? Choose items from page 11, exercise 7.

assemble (a skateboard) = fit the parts (of a skateboard) together

Listening 2 🔊 16 Listen and complete the checklist.

write: *1 mm*; say *one millimetre* or *one mil*
write: *5 mm*; say *five millimetres* or *five mil*
(Stress the underlined syllable)
size M5 = 5 mm

	Size	Quantity
spanner	____ mm	1
nuts	____ mm	____
bolts	M____	____

Speaking 3 Work in pairs. Make dialogues with your partner.

1 bolts / 10 mm / 50
2 washers / M6 / 60
3 screws / 24 mm / 100
4 nuts / 36 mm / 75
5 bolts / M16 / 60
6 nails / 30 mil / 80

Example:
Customer: I need some bolts, please.
Shopkeeper: What size?
Customer: 10 mm.
Shopkeeper: How many?
Customer: Fifty, please.

Task 4 How do you assemble a skateboard? Put these diagrams in order.

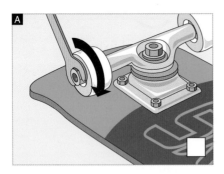
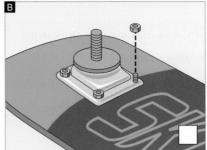
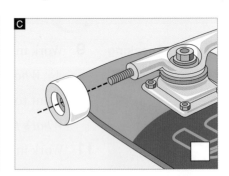
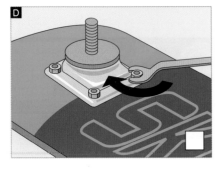
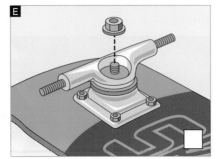
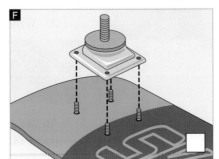

Reading   **5**   Read this instruction manual and check your answers to 4.

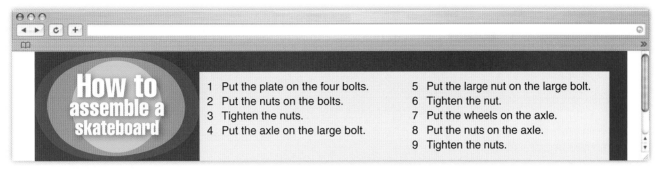

How to assemble a skateboard

1 Put the plate on the four bolts.
2 Put the nuts on the bolts.
3 Tighten the nuts.
4 Put the axle on the large bolt.
5 Put the large nut on the large bolt.
6 Tighten the nut.
7 Put the wheels on the axle.
8 Put the nuts on the axle.
9 Tighten the nuts.

Language   **6**   Complete the table. Use the sentences from 5. Leave some spaces blank.

Verb (action)	Object (thing)	Location (place)
1 Put	the plate	on the four bolts.
2 Put		
3	the nuts.	
4 Put		
5 Put		
6		
7		
8		
9		

Vocabulary   **7**   ▶ 🔊 17   Listen and repeat.

loosen … pull … push … put … take … tighten

8   Complete the instructions. Use the words from 7.

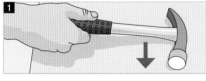

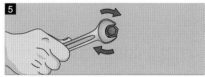

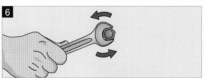

1 _____ the hammer on the table.
2 _____ the hammer off the table.
3 _____ the lever.
4 _____ the lever.
5 _____ the nut.
6 _____ the nut.

9   Complete the table.

Verb	Opposite
put (on)	(1) (off)
tighten	(2)
push	(3)

3 Ordering

Listening 1 🎧 18 Listen to this voice mail message and complete the notes.

Phone call from
Name: _Ben_   Phone number: _00 44_
Message: _Customer needs some skateboard parts. Please call him back._

2 🎧 19 Listen and correct the mistakes in these names and numbers.

write: 55; say: *five five* or *double 5*.
write: 0; say: *zero* or *oh*.

1	Abdel Monem Waheed 00 202 47832
2	José Fernandez Luis 00 34 912 838 990
3	Adel Al-Mansour 00 971 2 605 8843
4	Nikolay Kuznetsov 00 7 495 900 22 77

Speaking 3 Work in pairs. Choose words from this unit (e.g. *screwdriver*) and dictate them to your partner.

4 Work in pairs. Leave phone messages.

Student A. Turn to page 112.

Student B:
1 Leave phone messages for Student A. Use the business cards below. Spell out the name of the person and the company.

Example:
Hello. This is John West. That's W-E-S-T. Manager of Kesko. That's K-E-S-K-O. My phone number is 00 44 1224 867 4490. Please call me back.

STELLA MARITIMA

Pepino Turi
Engineer
00 39 06 625 500
①

NIKOMATIC
Kazuo Suzuki
Technician
00 81 3 3388 5124
②

Komet
Stefan Gross
Designer
00 49 711 845 8833
③

2 Change roles. Listen to Student A and make notes like this:

Call from John West, Manager
Company: Kesko
Phone number: 00 44 1224 867 4490
Please call him back.

Task 5 Work in pairs. Order goods on the phone.

Student A. Turn to page 112.

Student B:

1 You are a customer. You want to buy the items circled in red. Telephone Student A (the sales person) and order the items.

Item	Colour			Size			Quantity		
Helmet	red	white	(blue)	large	(medium)	small	(1)	2	3
Deck	(red)	yellow	blue	(large)	medium	small	(1)	2	3
Pad	black	(brown)	green	large	(medium)	small	2	(4)	6

Begin:
A: *Hello. I need to buy some things for my skateboard.*
B: *OK. What do you need?*
A: *I need a helmet.*

USEFUL PHRASES

What size/How many/What colour do you need?
What's your name? Please spell that.
What's your phone number?

2 Change roles. You are the sales person. Ask Student A (the customer) what they want to buy.

3 When you have both finished, you can circle new items and phone up to order them.

Social English 6 ▶ 🔊 20 Listen and then introduce yourself and your partner to other students.

A: *I'm Luis. I'm a student. And this is Paulo. He's a student, too.*
B: *Hello, Luis. Hello, Paulo. Nice to meet you.*

Review Unit A

1 Rewrite these statements as questions.

1 The machine's on.
   Is the machine on?
2 The switches are off.
   _____
3 Roberto's in London.
   _____
4 They're IT technicians.
   _____
5 He's a student.
   _____
6 She's Polish.
   _____

2 Answer the questions in the negative. Then make a positive statement.

1 Is it Sunday today? (Monday)
   No, it isn't Sunday today. It's Monday.
2 Is the power on? (off)
   _____
3 Are you Peter? (John)
   _____
4 Are they from Berlin? (Bonn)
   _____
5 Is she a technician? (engineer)
   _____
6 Is he an electrician? (builder)
   _____

3 Rewrite these sentences using contractions where possible.

1 My name is Jamal and I am from Jordan.
   My name's Jamal and I'm from Jordan.
2 This is Jean. He is French, but he is not from Paris.
   _____
3 This is Frieda. She is from Rome, but she is not Italian.
   _____
4 Look at the switch. It is down, but the power is not on.
   _____
5 These are the wrong items. They are not bolts. They are screws.
   _____
6 What is this tool called? What are these called?
   _____

4 Complete the questions and answers with the words in the box. You can use the words more than once.

am   are   do   does   is

1 Where _are_ you from?
2 What _____ you do?
3 Excuse me. _____ you Ian?
4 What _____ he do?
5 What _____ his name?
6 Excuse me. _____ they from France?

a) No, my name _____ Jan.
b) I _____ an IT technician.
c) His name _____ Peter.
d) No, they _____ from Germany.
e) I _am_ from Denmark.
f) He _____ a marine engineer.

5 Match the questions with the answers in 4.

6 Work in pairs. Practise the questions and answers in 4. Use contractions.

7 Look at the pictures in Units 1 and 2. Work in pairs. Make questions and answers about the pictures.

What's this/that called? What are these/those called?
It's/They're called … .

8 Look at this drawer for 15 seconds. Then close the book and list everything in the drawer.

Begin: 3 screws, …

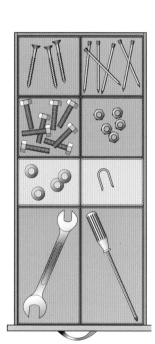

9 Draw a line from each word to its opposite.

on   stand   large   in   stop   off
left   small   up   sit   right
open   out   closed
start   tighten   loosen   down

10 Choose the correct way to read out these numbers.

1 Room 101
   a) one hundred and one
   b) one oh one

2 Height: 8850 metres
   a) eight thousand eight hundred and fifty
   b) double eight five oh

3 Tel: 74 77 88
   a) seventy-four seventy-seven eighty-eight
   b) seven four double seven double eight

4 Voltage: 109,845 V
   a) One hundred and nine point eight four five
   b) One hundred and nine thousand eight hundred and forty-five

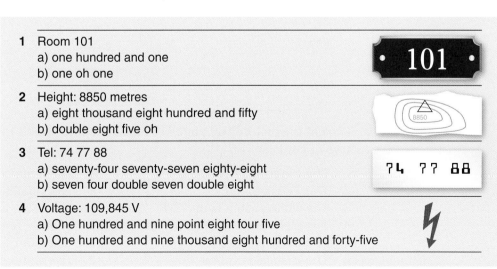

Review Unit A

11 Work in pairs. Solve this puzzle. Write a sentence of eight words.

pea	are		see	tea	eye		why	oh	you	are
P	A				S	E				

			eye						are	why				why
E	N	G	L	I	S	H	E	V	E			D	A	

eye	tea				are	eye		
W		H	A	F			E	N	D

12 Complete the dialogue with the question words in the box.

How   What   What's

1 ● _____ do you need?
  ○ *Some bolts, please.*
2 ● _____ many do you need?
  ○ *Forty, please.*
3 ● _____ size?
  ○ *10 mm, please.*
4 ● _____ colour? Black or silver?
  ○ *Black, please.*
5 ● _____ your name?
  ○ *John Martins.*
6 ● _____ your phone number?
  ○ *It's 00 30 438 9981.*

13 Say the dates and times. Use the 12-hour clock.

1 `WED 10/04/07  13.40`
2 `FRI 13/11/09  07.55`
3 `MON 03/09/10  11.05`
4 `WED 29/01/11  21.32`

Example: 1 Wednesday, the tenth of April 2007 at 1.40 pm.

14 Complete the number sequences with your partner.

a) 1, 2, 3, 5, 7, _____, _____, _____
b) 1, 1, 2, 3, 5, 8, _____, _____, _____
c) 2, 5, 10, 17, 26, _____, _____, _____
d) 0, 1, 10, 11, 100, 101, _____, _____, _____

15 Write these numbers and units in words.

1 5 km *five kilometres*
2 250 kg _____
3 €1015 _____
4 110 V _____
5 0°C _____
6 13 mm _____

16 Look at the pictures on page 113 for 15 seconds. Don't look again. Are these true (T) or false (F)?

1 The window is open. T/F
2 The TV is on. T/F
3 The white switch is up. T/F
4 The black switch is down. T/F
5 The circle is blue. T/F
6 The triangle is yellow. T/F
7 The large helmet is green. T/F
8 The small helmet is red. T/F
9 The cable is under the table. T/F
10 The car goes left. T/F
11 The letter is B. T/F
12 The number is 14. T/F

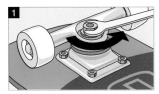
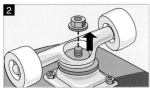
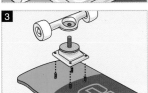

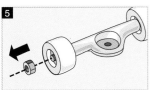
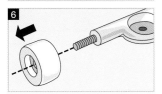

17 Complete the instructions for these pictures. Use SOME of the words in the box.

| loosen | off | on | put | take | tighten | use |

How to take the truck off the skateboard

Step 1: (a) _____ the large nut. (b) _____ the large spanner.

Step 2: (c) _____ the large nut (d) _____ the bolt.

Step 3: (e) _____ the truck (f) _____ the bolts.

How to take the wheels off the truck

Step 4: (g) _____ the small nuts. (h) _____ the small spanner.

Step 5: (i) _____ the small nuts (j) _____ the axle.

Step 6: (k) _____ the wheels (l) _____ the axle.

18 Put the words in the instructions in the correct order.

1 screws the tighten
  Tighten the screws.

2 the large hammer use
  _____

3 take off the car the old wheel
  _____

4 the new wheel put on the car
  _____

5 into the wood hammer the nails
  _____

6 through the holes the bolts push
  _____

Project **19** Find the meaning of the words *plate*, *truck* and *axle* for different technical fields, and write the results in a table.

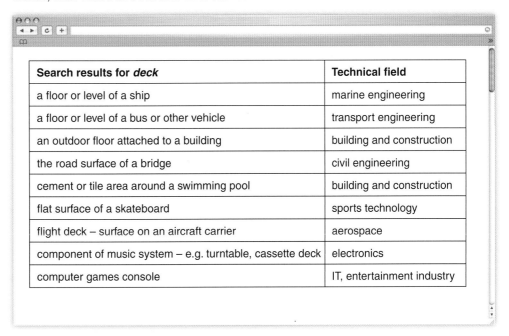

Search results for *deck*	Technical field
a floor or level of a ship	marine engineering
a floor or level of a bus or other vehicle	transport engineering
an outdoor floor attached to a building	building and construction
the road surface of a bridge	civil engineering
cement or tile area around a swimming pool	building and construction
flat surface of a skateboard	sports technology
flight deck – surface on an aircraft carrier	aerospace
component of music system – e.g. turntable, cassette deck	electronics
computer games console	IT, entertainment industry

3 Parts (2)

1 Tools

Start here 1 🔊 21 Listen and complete the TV advert.

This is the new Multi Tool!

Use it at home. Use it on the building site. Use it when you travel.
It has a (1) _____ and a pair of (2) _____.
It also has a (3) _____, a (4) _____ and a
(5) _____ _____.
The Multi Tool has everything you need! Only £29.99. Buy one now!

Listening 2 🔊 22 Listen and complete the dialogue with the words in the box.

do   does   doesn't   have

pliers and *scissors* are always plural
say: *I need some scissors*, or
I need a pair of scissors.

- Do you (1) _____ a Multi Tool?
○ Yes, I (2) _____.
- Does the Multi Tool (3) _____ a hammer?
○ Yes, it (4) _____.
- Does it (5) _____ a pair of scissors?
○ No, it (6) _____.

3 🔊 23 Listen and repeat.

a pair of pliers … a pair of scissors … a blade … a can opener …
a bottle opener … a screwdriver

Language

Do	you		have	a Multi Tool?	Yes, I do. / No, I don't.
Does	the Multi Tool			a hammer?	Yes, it does. / No, it doesn't.
	The Multi Tool	doesn't / does not	have	a hammer.	

4 Work in pairs. Practise the dialogue.

A: *Does Pedro have a Multi Tool?*   Bob / you / we
B: *Yes, he does.*   he / I / we
A: *Does it have a ruler?*   chisel / saw / spanner / screwdriver
B: *No, it doesn't.*   yes / no
A: *Does it have a pair of pliers?*   hammer / scissors / opener / blade
B: *Yes, it does.*   no / yes

5 Work in pairs. Design a Multi Tool for your work.

Reading   **6**   Complete the labels for this tool.

7   Read this product review and check your answers to 6.

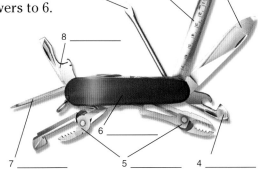

1 <u>screwdriver</u>   2 _____   3 _____
8 _____
6 _____
7 _____   5 _____   4 _____

Product review: Survival Tool
This tool has a ruler, a screwdriver, a pick and a blade. It also has two openers.
One opens cans. The other opens bottles. It has two wrenches.
It doesn't have a saw. And it doesn't have a hammer, because the tool is too small.
It has a plastic cover. The cover comes in three colours: black, blue or red.

Speaking   **8**   Ask and answer questions about the Survival Tool and the Multi Tool. Use the words in the box.

BrE *spanner*, AmE *wrench*

| blade   can opener   cover   hammer   pair / pliers   pair / scissors   ruler   wrenches |

A: Does the Survival Tool/Multi Tool have … ?
B: Yes, it does. / No, it doesn't.

Vocabulary   **9**   Match the parts to the words.

handle, shaft, head, blade, jaws

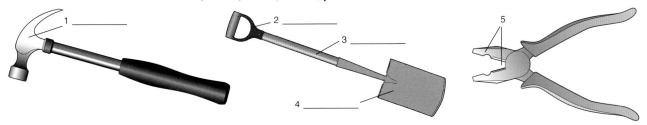

10   Draw some tools that you use in your work. Label some parts with words from 9. Then make sentences.

This is a … . It has a handle, a head and two jaws.

Speaking   **11**   Compare these three products.

Product comparison			
	Survival Tool	Multi Tool	Key Tool
knife blade	•	•	•
saw		•	•
screwdriver	•	•	•
bottle opener	•	•	•
can opener	•	•	
ruler	•		
pick	•		
wrench	•		
hammer		•	
pliers		•	•

1   The Key Tool has a screwdriver, but it doesn't have a wrench.
2   The Survival Tool has a ruler, but the Multi Tool doesn't.

Writing   **12**   Write a short comparison of the three products in 11.

2 Functions

Start here 1 Match the words with the pictures.

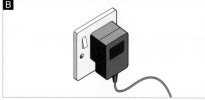

Electrical power sources
1 mains electricity + AC adapter
2 solar power
3 dynamo
4 batteries

Reading 2 Label the photos of the emergency radio below with the words in the box.

alarm   antenna   clock   compass   handle   thermometer   torch

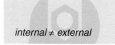

3 ▶️ 24 Listen and repeat.

handle ... thermometer ... torch ... alarm ... clock ... compass ... antenna

4 Read the description and check your labels.

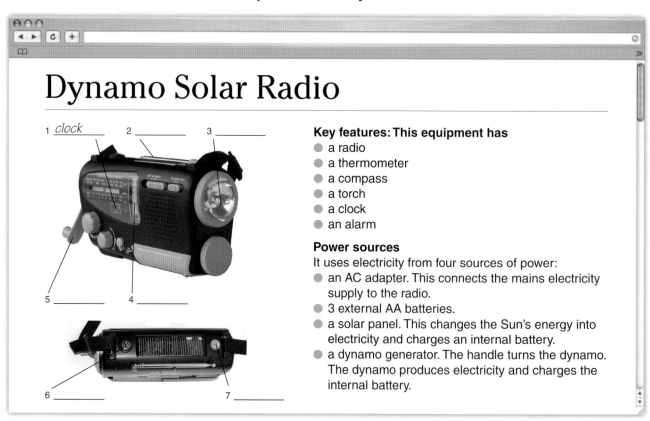

Dynamo Solar Radio

1 *clock* 2 _____ 3 _____

Key features: This equipment has
- a radio
- a thermometer
- a compass
- a torch
- a clock
- an alarm

Power sources
It uses electricity from four sources of power:
- an AC adapter. This connects the mains electricity supply to the radio.
- 3 external AA batteries.
- a solar panel. This changes the Sun's energy into electricity and charges an internal battery.
- a dynamo generator. The handle turns the dynamo. The dynamo produces electricity and charges the internal battery.

5 Explain the function of these parts.

1 the AC adapter   3 the dynamo
2 the handle   4 the solar panel

22   3 Parts (2)

6 Match the parts with their functions.

Part	Function
1 thermometer	a) shine a light
2 compass	b) make electricity
3 torch	c) turn the dynamo
4 clock	d) tell the time
5 alarm	e) find North
6 solar panel	f) receive radio signals
7 handle	g) measure temperature
8 antenna	h) make a loud noise

7 Make sentences from the parts and functions in 6.

Example: 1 The thermometer measures temperature.

Language

	It		measure	s	temperature.	
Does	it		measure		temperature?	Yes, it does. / No, it doesn't.
	It	does not / doesn't	measure		speed.	

Speaking

8 Work in pairs. Make questions and answers, using the words from 6.

A: *Does a thermometer measure time?*
B: *No, it doesn't. It measures temperature.*

9 Match the pictures with the verbs in the box.

cut   drive in   grip   loosen   tighten

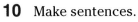

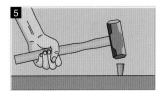
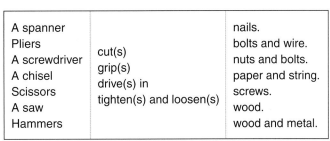

10 Make sentences.

A spanner Pliers A screwdriver A chisel Scissors A saw Hammers	cut(s) grip(s) drive(s) in tighten(s) and loosen(s)	nails. bolts and wire. nuts and bolts. paper and string. screws. wood. wood and metal.

Social English

11 Make a list of job titles useful to you. Use a dictionary.

Examples: marine technologist, computer operator, automotive engineer, architectural technician

12 Find out about other students in your class.

A: *What do you do?*
B: *I'm a/an ... . (student/builder/mechanic ...)*
A: *Where do you study/work?*
B: *I study/work at ... . (name of school/college/company ...)*
A: *What does ... do?*
B: *He/She's a/an ... . He/She works at ... .*

3 Locations

Start here 1 ▶ 🔘 25 Listen to this computer lesson. Complete the dialogue with the words in the box.

| at bottom on left right top |

- OK, now put the cursor on the START button.
○ Where's the START button?
- It's _____ the _____. On the _____. Do you see it?
○ Yes. Is that it?
- Yes, that's correct. ... Now, move the cursor up to the CLOSE button.
○ Where's that?
- It's an X. It's _____ the _____. At the _____.
○ Is that it?
- Yes, that's it. Now click.

Vocabulary 2 Match the TV monitors with their locations.

middle = centre
BrE centre, AmE center

1 bottom left ___
2 bottom right ___
3 centre bottom ___
4 centre left ___
5 centre right ___
6 centre top ___
7 top left ___
8 top right ___
9 centre ___

Language
in in the middle, in the centre
at at the top, at the bottom
on on the left, on the right

Reading 3 Correct this description. There are six mistakes in location.

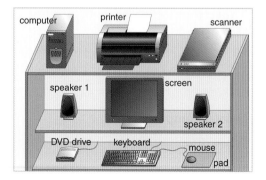

Here is one way to set up your computer station. Put your screen in the centre of the system. Then put one speaker in the centre on the left, and put the other speaker in the centre on the right. Put the scanner at the top on the left, and put the computer at the top on the right. Then put the DVD drive at the top in the middle and put the printer at the bottom on the left. Finally, put the keyboard at the bottom on the right, and put the mouse at the bottom in the centre.

24 3 | Parts (2)

Language

4 Look again at the computer station in 3. Are these statements true or false?

1. The computer is *at the top, on the left*. T/F
2. The computer is *above* speaker 1. T/F
3. The computer is *to the left of* the printer. T/F

5 Look at the diagram. When do we use *ON the left* and when do we use *TO the left OF*?

6 Complete the sentences about the computer station in 3 with the words in the box.

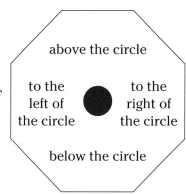

| above | at | below | of | in | on | to |

1. The printer is _____ the top, _____ the middle.
   The scanner is _____ the right _____ the printer.
   The screen is _____ the printer.
2. The mouse is _____ the bottom, _____ the right.
   The keyboard is _____ the left _____ the mouse.
   Speaker 2 is _____ the mouse.

7 Look again at the computer station in 3. Make sentences about the location of:

1 the mouse   2 the DVD drive   3 the scanner   4 the screen

Task

8 Work in pairs. Student A. Turn to page 113.

Student B:
1. Answer Student A's questions. Use phrases from exercise 6.
2. Ask Student A where these items are and write them in their correct locations: *mouse pads, scanners, CD-ROMS, adapter, printers, amplifiers, TV*.

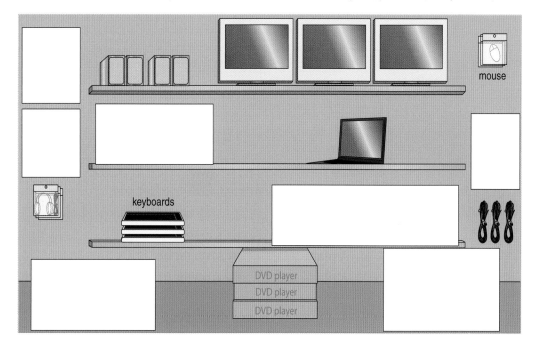

shelf (singular); *shelves* (plural)

4 Movement

1 Directions

Start here

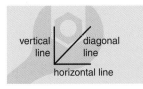

1 Label the jetpack man's movements with the words in the box.

backwards   down   forwards   sideways   up

2 Work in pairs. Which directions can planes and helicopters fly? Tick the boxes.

Direction	Plane	Helicopter
forwards		
backwards		
up and down		
sideways		

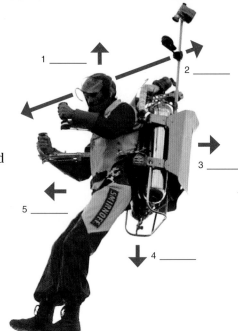

1 _____
2 _____
3 _____
4 _____
5 _____

Reading

3 Read the text. Check your answers to 2.

> Passenger planes can fly forwards, and can turn to the left and to the right. But they cannot fly backwards or sideways. They can fly diagonally up and down, but they cannot fly straight up or straight down.
> Helicopters can fly forwards, straight up and down, sideways and backwards.
> Planes and helicopters can both rotate. Planes and helicopters can rotate on their horizontal axis. Helicopters can also rotate on their vertical axis.

Language

It They	can	fly	sideways.
	can't/cannot		

Can	it they		fly	sideways?	Yes, it can. / No, it can't Yes, they can. / No, they can't

4 Complete these sentences with *can* or *can't*.

1 A helicopter _____ fly sideways, but a plane _____.
2 A plane _____ fly sideways, but it _____ fly forwards.
3 A plane _____ fly straight up, but a helicopter _____.
4 A plane _____ fly straight up, but it _____ fly diagonally.

Speaking

5 Work in pairs. Practise dialogues.

helicopter(s) / rocket(s) / plane(s) / fly sideways / fly straight up / fly diagonally up / rotate

A: *Can a plane fly forwards?*   B: Yes, it can.
A: *Can it fly backwards?*   B: No, it can't.

Task   **6**   Work in pairs. Follow the instructions and answer the questions.

Close your fist and hold your arm out straight in front of you.

1. Think of your wrist. (Don't move it). How many directions can it move in? One, two, three or four?
2. Think of your shoulder. (Don't move it). How many directions can it move in? One, two, three or four?
3. Think of your elbow. (Don't move it). How many directions can it move in? One, two, three or four?

Reading   **7**   Read the text. Check your answers to 6.

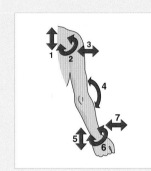

The human arm can move in seven different directions. The arm has three pivots: the wrist, the elbow and the shoulder. The wrist can move in three different directions. At the wrist, the hand can move up and down about 90°, it can move from side to side about 70°, and it can rotate about 180°. The shoulder can move in the same three directions, but different angles. It can rotate about 20°. The elbow can only move in one direction. At the elbow, the forearm can only move up and down. It cannot move sideways or rotate.

8   Match each movement in the diagram in 7 with a word or phrase from the box.

    rotate   move sideways   move up and down

Listening   **9**   ▶ 26   Listen and choose the correct answers.

1. a) 19°   b) 90°   3   a) 17°   b) 70°
2. a) 14°   b) 40°   4   a) 118°   b) 180°

Task   **10**   Work in groups. Look at the diagram in 11 and answer these questions.

1. How many directions can this robot arm move?
2. Which part of the robot arm has different movements from the human arm. Is it: a) the shoulder?   b) the wrist?   c) the elbow?

Language   **11**   Complete the text about the robot arm with the words in the box.

    can   can't   has   is   isn't

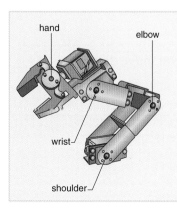

This robot arm (1) _____ like a human arm. It (2) _____ a 'wrist', an 'elbow' and a 'shoulder'.
The 'wrist' (3) _____ like the human wrist. It (4) _____ three movements. It (5) _____ rotate. It (6) _____ move from side to side. It (7) _____ move up and down.
The 'elbow' (8) _____ like the human elbow. It (9) _____ one movement. It (10) _____ move up and down.
The 'shoulder' (11) _____ like the human shoulder, because it only (12) _____ two movements. It (13) _____ rotate, and it (14) _____ move up and down. But it (15) _____ move sideways.

Movement   **4**

2 Instructions

Start here 1 Try this quiz. Choose the correct answer.

km/h = kilometres per hour (used by most countries)
mph = miles per hour (used in some countries, including the US and UK)
m/s = metres per second
rpm = revolutions per minute; 1 revolution = 1 rotation of 360°

What are the speeds?

1. Rotation of a fast CD-ROM?
   a) 98,000 rpm   b) 9800 rpm
2. The speed of sound?
   a) 746 km/h (464 mph)   b) 1200 km/h (746 mph)
3. The maximum speed on land?
   a) 1228 km/h (763 mph)   b) 1228 mph (1976 km/h)
4. The maximum speed on water?
   a) 154 m/s   b) 154 mph   c) 154 km/s
5. The rotation of the Earth?
   a) 1000 mph (1609 km/h)   b) 1000 km/h (621 mph)
6. The Earth moving around the Sun?
   a) 67,000 mph (107,825 km/h)   b) 67,000 km/h (41,631 mph)

Listening 2 🔊 27 Listen and check your answers to 1.

3 Work in pairs. Write down some speeds. Dictate them to your partner.

Vocabulary 4 Label the diagram with the words in the box.

antenna   handle   joysticks   slider   switch

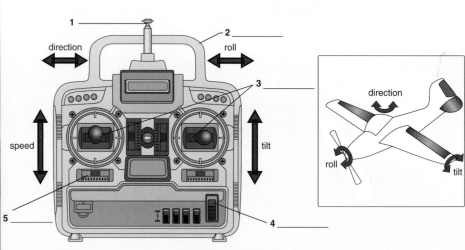

Task 5 Work in groups. What do you think the plane does when you move these controls?

Look at the joystick on the left.
1 Push it up (away from you). Pull it down (towards you). What happens?
2 Push it to the left. Push it to the right. Now what happens?

Look at the joystick on the right.
3 Push it up. Pull it down. What happens?
4 Push it to the left. Push it to the right. Now what happens?

4 Movement

Reading   **6**   Read the manual. Check your answers to 4 and 5.

Remote control transmitter for model plane

User manual

Look at the diagram of the transmitter. There are two joysticks. One is on the left. This is the left-hand (LH) stick. The other is on the right. This is the right-hand (RH) stick.

Now look at the LH joystick. This controls the speed and the direction of the plane. Push the LH stick up (away from you) and the plane accelerates. Pull (it) down (towards you) and the plane slows down. Slide
5 the stick to the left and the plane turns left. Slide it to the right and (it) turns right.

Now look at the RH joystick. This controls the roll and the tilt of the plane. Push the RH stick up (away from you) and the plane descends (or goes down). Pull it down (towards you) and the plane ascends (or goes up). Slide the stick to the left and the plane rolls to the left. Slide it to the right and it rolls to the right.

7   Which words in the text do these pronouns refer to?

1   it (line 4)   a) direction   b) plane   c) LH stick
2   it (line 5)   a) plane   b) LH stick   c) right

8   Match your actions with the plane's actions.

Your action	The plane's action
1 Move the LH stick up.	a) The plane goes to the left.
2 Pull the LH stick down.	b) The plane goes faster.
3 Move the LH stick to the left.	c) The plane goes down.
4 Move the LH stick to the right.	d) The plane goes more slowly.
5 Move the RH stick up.	e) The plane rolls to the left.
6 Pull the RH stick down.	f) The plane goes up.
7 Move the RH stick to the left.	g) The plane rolls to the right.
8 Move the RH stick to the right.	h) The plane moves to the right.

ascend ≠ descend

Speaking   **9**   Work in pairs. Make dialogues with the information from the table in 8.

A: *Can the plane fly to the left?*
B: *Yes, it can. You move the left-hand stick to the left.*

Social English   **10**   Work in pairs. Find out what your partner can and can't do.

A: *Can you swim?*   B: *Yes, I can. Can you?*
A: *Yes, I can. Can you sail a boat?*   B: *No, I can't.*

3 Actions

Start here 1 Look at the diagrams and answer the questions.

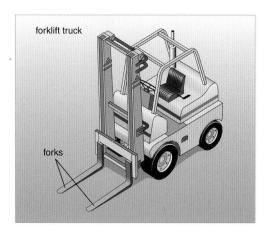
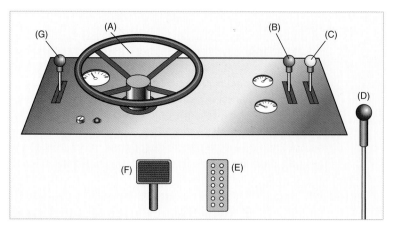

1 How many pedals does it have?
2 How many levers does it have?
3 Is the steering wheel on the left or on the right?

Reading 2 Read the manual. Write the letters (A–G) from the diagram next to the controls.

reverse = go backwards

> In the diagram, you can see the controls of the forklift truck. On the left is a lever. This is the direction lever (1 _____). Push this lever forwards, and the truck moves forwards. Pull it backwards, and the truck reverses. Next you can see the steering wheel (2 _____). This turns the truck to the left and right. At the top, on the right, you can see two levers. Push the left-hand lever (3 _____) forwards, and the fork moves up. Pull it back, and the fork moves down. Push the right-hand lever (4 _____) forwards, and the fork tilts up. Pull it back, and the fork tilts down. At the bottom, on the right, you can see a lever. This is the parking brake (5 _____). At the bottom, you can see two pedals. The LH pedal is the brake (6 _____). The RH pedal is the accelerator (7 _____).

3 Describe these movements of the truck. Use words from the manual.

Example: A. The fork tilts down.

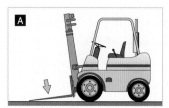

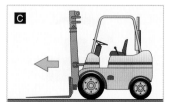
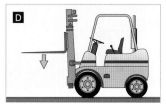

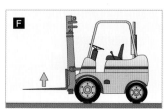

Speaking 4 Work in pairs. Have a driving lesson.

Student A: You are the driving instructor. Give instructions.
Student B: You are learning to drive. Follow the instructions. Act them out.

Drive forwards. Reverse. Go slowly. Go faster. Slow down. Stop! Turn left. Turn right. Reverse to the left. Reverse to the right. Turn round. Do a U-turn. To the left. To the right.

turn round = do a U-turn

Writing 5 Write a short set of instructions for one of these jobs. Draw a diagram.

1 How to park a car.
2 How to dock a small sailing boat.
3 (Choose your own job.)

6 Write full sentences from these notes. Use **when** and **you** and add **the** and punctuation.

1 pull lever C backwards → fork tilts down
2 push lever B forwards → fork moves up
3 turn steering wheel to the right → truck turns right
4 pull lever G backwards → truck reverses
5 press brake pedal → truck stops
6 press accelerator → truck goes faster

Example: 1 When you pull lever C backwards, the fork tilts down.

Task 7 Work in pairs. Have a driving lesson for the forklift truck.

Student A. Turn to p. 115.

Student B:
1 You're the driving instructor for the forklift truck. Student A is learning to drive the truck. Tell Student A to follow these instructions in the correct sequence.
2 Then change roles. Follow Student A's instructions and rearrange your pictures into the correct sequence.

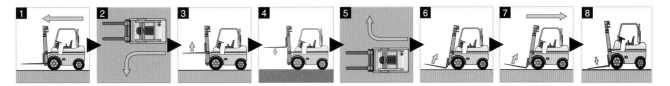

The correct sequence of the instructions is:

Review Unit B

1 Rewrite these statements as questions.

1. John has the spanners.   *Does John have the spanners?*
2. The students have a holiday tomorrow. _____
3. The Multi Tool has a screwdriver. _____
4. These bikes have strong brakes. _____
5. The radio has an internal battery. _____
6. Those houses have solar panels. _____

2 Answer the questions in the negative. Then make a positive statement.

1. Do you have a car? (motorbike)
   No, I don't have a car. I have a motorbike.
2. Does your brother have a DVD? (VCR)
   _____
3. Does the Multi Tool have scissors? (knife blade)
   _____
4. Do we have English today? (Science)
   _____
5. Does your radio have batteries? (dynamo)
   _____
6. Do the pliers have plastic handles? (metal handles)
   _____

3 Rewrite these sentences using contractions where possible.

1. The Multi Tool does not have a wrench. It is not very useful.
   The Multi Tool doesn't have a wrench. It isn't very useful.
2. We do not have an AC adapter. We can not switch on the computer.
   _____
3. I am a technician, but I do not have my tools here. I can not repair your TV.
   _____
4. The electricity is off, and we do not have any batteries. You can not use the radio now.
   _____

4 Give short answers.

1. Can you swim? (No) *No, I can't.*
2. Is he an IT technician? (No) _____
3. Does the DVD work now? (Yes) _____
4. Do your friends have tickets for the cinema? (No) _____
5. Are you a technology student? (Yes) _____
6. Does your radio have a solar panel? (No) _____
7. Are you a telecoms engineer? (No) _____
8. Can planes rotate on a horizontal axis? (Yes) _____

5 Complete the dialogue with the correct form of the verbs in brackets.

● *Look at my radio. Do you like it?*
○ Yes, it's great. What (1) _____ (do) that handle (2) _____ (do)?
● *It (3) _____ (turn) a dynamo. The dynamo (4) _____ (produce) electricity for the radio.*
○ What are those, at the top?
● *They're solar panels. They (5) _____ (charge) the internal battery on a sunny day.*
○ Can the radio also (6) _____ (use) mains electricity?
● *Yes, it can. And it also (7) _____ (use) AA external batteries.*
○ So your radio (8) _____ (have) four power sources!
● *That's right.*

6 Label the parts with the words in the box.

blade/blades   handle/handles   head   jaws   shaft

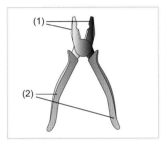

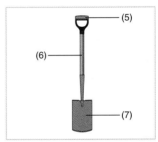
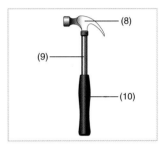

7 Describe the tools in 6.

Example: 1. A pair of pliers has two handles and two jaws.

8 Look at this toolboard for 15 seconds. Then close the book and list all the tools.

Begin: Five screwdrivers. They're at the top, on the left.

9 Look again at the toolboard on the right. Make sentences with the words in the box.

above   below   to the left of
to the right of

Example: The screwdrivers are to the left of the spanners and above the hammer.

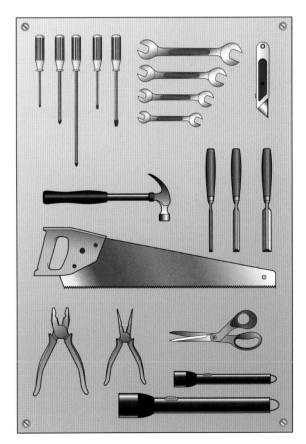

Review Unit B

10 Guess the device from its description.

1. This item covers the head and protects it. Skateboarders use it.
2. This tool has handles and jaws. It can grip nuts and bolts. It pulls nails out.
3. This equipment converts (or changes) sunlight into electricity.
4. You rotate these items onto bolts. You tighten them with a spanner.
5. This item receives radio and TV signals. You can see it on a house or car.
6. This equipment produces electricity when it rotates.

Notice the spelling change:
study → studies

11 Complete these questions and answers with the words in the box.

| am are come/comes do does is study/studies work/works |

1. Where _are_ you from?
2. What _____ you do?
3. Where _____ you study?
4. What _____ your subject?
5. Where _____ Elli come from?
6. _____ she a student, too?
7. What _____ she do?
8. Where _____ she work?

a) She _____ at Vodafone.
b) I _____ a student.
c) She _____ a technician.
d) She _____ from Finland.
e) I _am_ from Japan.
f) I _____ at the Technical College.
g) I _____ telecoms engineering.
h) No, she _____ not.

12 Match the questions with the answers in 11.

13 Work in pairs. Practise the questions and answers in 11. Use contractions where possible.

Example: 1–e   A: Where are you from?   B: I'm from Japan.

14 Cross out the silent letters in the words below. Say the words.

1. knife
2. build
3. building
4. scissors
5. wrist
6. ascend
7. descend
8. right
9. tighten

Example: ~w~rench

15 Label the controls with the words in the box.

| button display key lever pedal slider switch wheel |

16 Put *a*, *an*, *some* or *a pair of* before each item.

To buy:

_____ printer _____ AC adapter _____ speakers _____ keyboard _____ amplifier _____ headphones _____ earphones _____ nuts _____ bolts _____ pliers

17 Make positive and negative statements.

1 this opener … open bottles ✓ open tins ✗
2 these wrenches … tighten the M12 bolts ✗ loosen the M5 nuts ✓
3 that antenna … receive radio signals ✓ transmit them ✗
4 a rocket … fly straight up ✓ reverse ✗
5 passenger planes … fly sideways ✗ turn left and right ✓
6 I … drive a car ✓ operate a forklift truck ✗

Example: 1 This opener can open bottles, but it can't open tins.

18 Follow the instructions.

Start at the red triangle. Move sideways three boxes to the right. Go diagonally up one box to the right. Move horizontally eight boxes to the left. Descend vertically three boxes. Go diagonally up two boxes to the right. Move diagonally down two boxes to the right. Where are you?

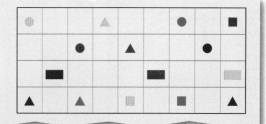

19 Match pictures with the instructions below.

1 Fly diagonally down.   6 Reverse to the left.
2 Fly forward.   7 Turn left.
3 Fly straight up.   8 Rotate on a horizontal axis.
4 Reverse.   9 Turn right.
5 Rotate on a vertical axis.   10 Reverse to the right.

Projects

20 Make a list of job titles in your industry.

Example: Construction Industry: structural engineer, quantity surveyor, site manager, architectural technician, etc.

21 What do these word parts mean? Find other words with the same part.

Word part	Meaning of word part	Example of word	Meaning of word
multi-		1 multimedia 2	1 2
therm-		1 thermometer 2	1 2
kilo-		1 kilometre 2	1 2

Review Unit B

5 Flow

1 Heating system

Start here 1 Work in groups. Which way does the water flow in this system? Draw arrows to show the direction of the flow.

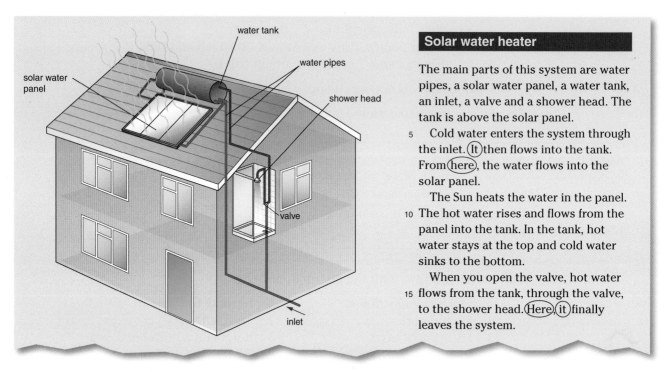

Solar water heater

The main parts of this system are water pipes, a solar water panel, a water tank, an inlet, a valve and a shower head. The tank is above the solar panel.

5 Cold water enters the system through the inlet. It then flows into the tank. From here, the water flows into the solar panel.

The Sun heats the water in the panel.
10 The hot water rises and flows from the panel into the tank. In the tank, hot water stays at the top and cold water sinks to the bottom.

When you open the valve, hot water
15 flows from the tank, through the valve, to the shower head. Here it finally leaves the system.

Reading 2 Read the text. Check the directions of your arrows in 1.

3 What do these words refer to?

1	It (line 6)	a) inlet	b) cold water	c) system
2	here (line 7)	a) tank	b) inlet	c) water
3	Here (line 16)	a) tank	b) valve	c) shower head
4	it (line 16)	a) shower head	b) valve	c) hot water

Example: 1 Cold water enters the system through the inlet. It then ... .
In line 6, *it* refers to *cold water*.

4 Draw the flow chart, putting these boxes into the correct order.

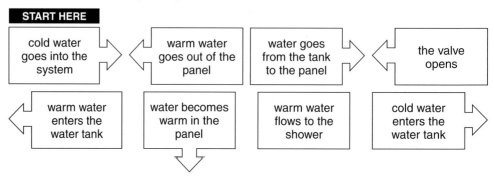

Language

The water	flow move	-s	into the tank. out of the tank.
	go pass	-es	through the pipes. around the solar panel. to the outlet. from the inlet.
The electron	-s	flow go	around the circuit. through the cables.

Vocabulary

5 Label the diagrams 1–6 with the prepositions in the box.

around   from   into   out of   through   to

6 Complete the table with the verbs in the box.

enter   leave   rise   sink

	up	(1)
go	down	(2)
	in/into	(3)
	out/out of	(4)

7 Complete the sentences with the correct form of verbs from the table in 6.

1 Water _____ the house through the inlet pipe.
2 Water _____ the solar panel through the outlet pipe.
3 When you heat the water in a tank, the hot water _____.
4 When you cool the air in a room, the cool air _____.

Task

8 Work in pairs. Explain your system to your partner.

Student A. Turn to page 114.

Student B:
1 Listen to Student A, and ask questions. Then draw a simple diagram of his/her system.
2 Explain your system to Student A.

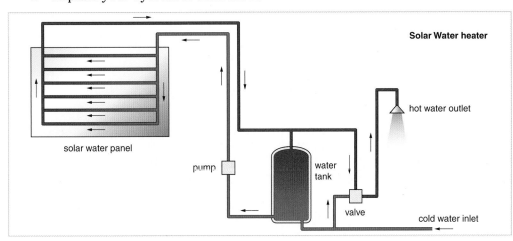

Writing

9 Write an explanation of your system.

2 Electrical circuit

Start here 1 Do you know these electrical symbols? Label the circuit diagram with the words in the box.

battery   conductor   fuse   lamp
negative   positive   switch   terminal

See the glossary of electrical symbols on page 109 for answers.

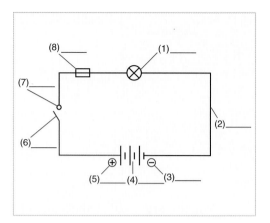

Listening 2 🔊 28 Listen and label the diagram with the words in the box.

battery   cables   controller   lamps   solar panel

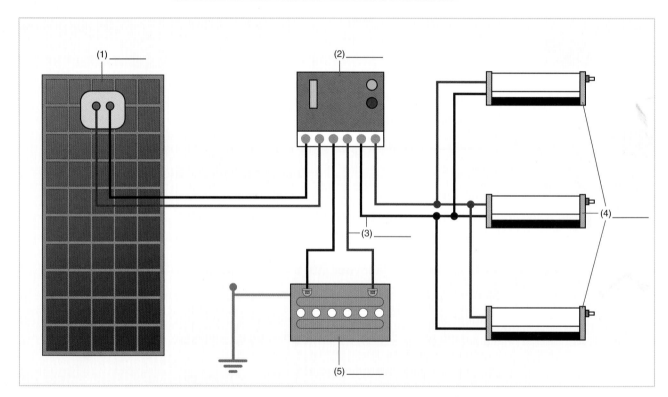

3 🔊 29 Listen and match the items with their specifications.

1 solar panel   a) 12V 8W
2 controller   b) DC
3 battery   c) 5A
4 lamps   d) 60W
5 electrical current e) 12V 100Ah

Task 4 Work in pairs. Look again at the diagram in 2. Where does the current flow in these three situations? Draw arrows.

Situation 1: The Sun shines. The lamps are on.
Situation 2: The Sun shines. The lamps are off.
Situation 3: The Sun doesn't shine. The lamps are on.

convert = change

5 Read the manual for the solar panel and check your answers to 4.

How does the solar power system work? The panel converts the Sun's energy into a DC electric current. The current flows to the controller. Then it can flow from the controller to the lamps. Or it can flow from the controller into the battery. The battery stores the electricity. The current can flow from the battery into the lamps through the controller.

If the Sun shines, the DC current can flow from the panel, through the controller and into the lamps. If the Sun doesn't shine, the current can flow from the battery, through the controller and into the lamps. If the lamps are off, the current can flow from the panel, through the controller, and into the battery.

The controller controls the flow of the current. If the battery is full, the controller stops the flow from the panel into the battery. If the battery is empty, the controller stops the flow from the battery into the lamps.

Language

If	the Sun		shine	-s	,	the current flows from the panel.
	the Sun	does not/doesn't	shine		,	the current flows from the battery.

If	the battery	is	full	,	the current doesn't flow into the battery.
	the lamps	are not/aren't	on	,	the current flows into the battery.

Task

6 Work in pairs. How do you think the controller below works? Make notes.

7 Complete the text explaining how the controller works. Choose the correct verb and use the correct form of the verb.

If the battery is full, switch A (1) _____ (open/<u>close</u>). Then the current (2) _____ (flow/not flow) from the panel to the battery. The controller short-circuits the panel.

If the battery is empty, switch B (3) _____ (open/close). Then the current (4) _____ (flow/not flow) from the battery to the lamp.

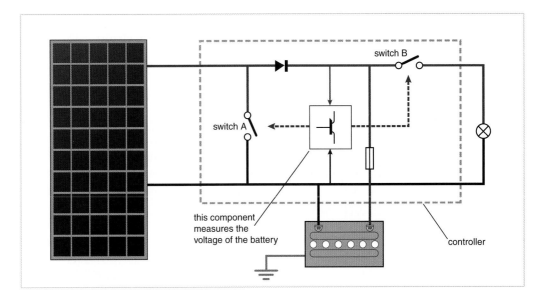

3 Cooling system

Start here 1 Try this quiz. Choose the correct answer.

What are the normal or average temperatures for these?

1. Water from a shower? a) 60°C (140°F) b) 80°C (176°F)
2. Food in a refrigerator? a) 4.5°F (−15°C) b) 40°F (4.5°C)
3. Food in a freezer? a) 0°C (32°F) b) −18°C (0°F)
4. Coldest air temperature ever? a) −89°C (−128°F) b) −20°C (−4°F)
5. Hottest air temperature ever? a) 156°F (70°C) b) 136°F (58°C)
6. Water in running car engine? a) 110°C (230°F) b) 45°C (110°F)

°F = °C * 9 / 5 + 32.
This converts Celsius to Fahrenheit.
°C = (°F − 32) * 5 / 9. This converts Fahrenheit to Celsius.

Listening 2 ▶ 30 Listen and check your answers.

Reading 3 Label the diagram with the words in the box.

bottom hose   engine   radiator   thermostat   top hose   water pump

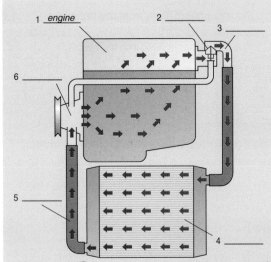

Car cooling system

The engine drives the water pump and the pump pushes cool water around the engine. This cools the engine. At the same time, the water becomes hot. The water in a hot engine is normally around 110°C.

5   The hot water then passes through the thermostat. (This) controls the temperature of the engine. From the thermostat, (it) flows through the top hose into the radiator. (Here), a fan cools the water, and the cool water sinks to the bottom of the radiator.

10   The cool water then leaves the radiator. (It) flows along the bottom hose, passes through the pump and enters the engine again.

4 Read the text. Check your answers to 3.

5 Which words in the text do these words refer to?

1. This (line 6)   a) hot water   b) thermostat   c) temperature
2. it (line 7)   a) engine   b) thermostat   c) water
3. Here (line 8)   a) top hose   b) radiator   c) fan
4. It (line 10)   a) water   b) radiator   c) bottom hose

Speaking 6 Make true sentences.

(1) The water pump	control(s)	the radiator to the engine.
(2) The thermostat	connect(s)	air onto the radiator.
(3) The two hoses	push(es)	the hot water from the engine.
(4) The radiator	cool(s)	water around the engine.
(5) The fan blades	flow(s)	to the bottom of the radiator.
(6) Cool water	rise(s)	the temperature of the water.
(7) Hot water	sink(s)	through the two hoses.
(8) Water	blow(s)	to the top of the engine.

Task 7 Work in groups. This is a simple way to cool a house in a hot country. How does it work? What happens at each stage (1–11)?

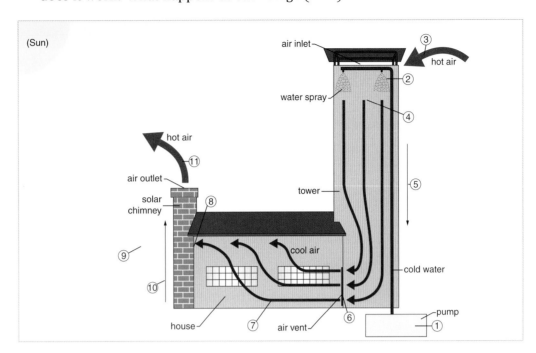

Writing 8 Complete this description of how the cooling system works with the verbs and prepositions in the box.

| cool | enter | flow | heat | leave | rise | sink |
| around | into | out of | through | to | | |

The pump pushes cold water _through_ the pipe _to_ the top of the tower (1).
Here, the water _leaves_ the pipe _through_ small holes. It's like a cold shower. (2).
Hot air _____ the tower _____ the air inlet (3).
The shower of cold water _____ the air (4). The cool air then _____ to the bottom of the tower (5).
The cool air _____ the house, (6) and then it _____ (7).
It _____ the house and _____ the solar chimney (8).
The Sun _____ the chimney, (9) and the hot air _____ (10).
The hot air finally _____ the chimney _____ the air outlet (11).

Social English 9 🔊 31 Listen and read.

Dan is an electronics student. He also works part-time in an electronics workshop.

- ● *I work in the electronics workshop every Thursday and Friday.*
- ○ When do you attend lectures?
- ● *Every Tuesday morning.*
- ○ What do you do on Tuesday afternoons?
- ● *I do my practical work then.*

on Mondays = every Monday
on Monday mornings = every Monday morning

10 Work in pairs. Practise the dialogue.

11 Work in pairs. Discuss your own weekly schedule.

6 Materials

1 Materials testing

Start here 1 Work in pairs. Read the instructions and answer the question.

- Look at the helmet and rope. What are they made of?
- Design tests for them. Use diagrams and the words in the box.

| break nylon polycarbonate pull stretch strike |

Listening 2 🔊 32 Listen and answer the questions.

1 What material is the rope made of?
   _____

2 What is the lecturer doing?
   _____

3 Is the rope breaking?
   _____

3 Listen again and complete the dialogue.

● I'm (1) _____ the rope. I'm (2) _____ it.
   Is it (3) _____?
○ No, it (4) _____.
● That's right. It (5) _____ _____.

Language This is the *present continuous* form of the verb. Use it to describe what is happening at the same time as you are speaking.

			'm / am	pull	-ing	the rope.
	The rope		isn't / is not	break	-ing.	
What	are	you		do	-ing?	
	Is	the rope		break		

Vocabulary 4 Match the actions with the verbs in the box.

| bend compress cut drop heat scratch stretch strike |

Language

5 The lecturer is testing other materials. Complete his description.

● Now I (1) *'m heating* (heat) this plastic to 100°C. Can you see?
It (2) _____ (not melt).
OK, now I (3) _____ (put) this helmet on the floor. And now the machine (4) _____ (drop) a 10 kg weight on it.
Right, now look at Dr Wilson. He (5) _____ (strike) the metal plate with a hammer. But the plate (6) _____ (not bend).
OK, now the jaws of the vice (7) _____ (compress) this plastic block. The block (8) _____ (not break).
Now Dr Wilson (9) _____ (hang) a weight of 500 kg from the ropes. But the ropes (10) _____ (not stretch).

Note the spelling changes:
strike → striking
drop → dropping
cut → cutting

Speaking

6 What are the people in the gym doing? Describe this picture using the words in the box.

| bend | cycle | hold | lift | pick up | pull | push | run | sit | stretch | touch |

7 Ask and answer questions about the picture in 6.

A: What's D doing? Is he pushing the bar up?
B: No, he isn't. He's pulling the bar down.

8 Work in pairs. Guess the sport from the mime.

Student A: do the actions.
Student B: guess what Student A is doing. Then change roles.

A: Watch me. (Mime a sport). What am I doing now?
B: Are you diving?
A: No, I'm not diving.
B: I know. You're swimming.
A: Yes, you're right. I'm swimming.

Materials | 6

2 Properties

Start here

1 Work in pairs. What are the most important properties of the materials in the box? Discuss with your partner.

> ceramic   concrete   diamond   fibreglass   graphite   steel

Example: You can't burn/melt/break/scratch/bend/cut it (easily).

Vocabulary

2 What are these made of? Match the photos with these materials.

> aluminium   ceramic   fibreglass   graphite   nylon
> polycarbonate   polystyrene   rubber   steel   titanium

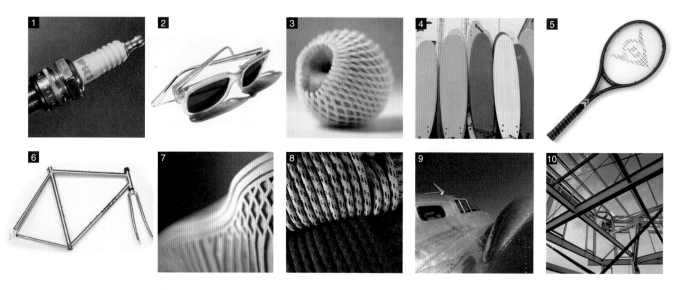

Speaking

3 Underline the stressed syllable.

1. ny lon
2. graph ite
3. ce ram ic
4. pol y car bon ate
5. al u min i um
6. pol y sty rene
7. ti ta ni um
8. fi bre glass

fibreglass (BrE) = *fiberglass* (AmE)
aluminium (BrE) = *aluminum* (AmE)

4 ▶ 🎧 33 Listen and check your answers to 3. Say the words with the correct stress.

Language

What	is / 's	this helmet	made of?	It	is / 's	made of	polycarbonate. nylon.
What	are / 're	those ropes		They	are / 're		

5 ▶ 🎧 34 Listen and repeat.

- What's this made of?
- It's made of ceramic.
- What are these made of?
- They're made of polycarbonate.

6 Work in pairs. Make similar questions and answers about the photos in 2.

Vocabulary

7 Match the sentences.

1. This material doesn't burn or melt if you heat it.
2. This material doesn't break if you strike it or drop it.
3. You can't bend this material.
4. This material doesn't corrode if you put it in water.
5. You can't scratch this material or cut it.

a) It's rigid.
b) It's hard.
c) It's tough.
d) It's heat-resistant.
e) It's corrosion-resistant.

8 Match the words with their opposites.

1. tough
2. hard
3. rigid
4. strong
5. light

a) soft
b) heavy
c) weak
d) brittle
e) flexible

Reading

9 Read the text and complete the table below.

This racing car is made from the latest hi-tech engineering materials. It's made from metals, alloys, ceramics, plastics and composites. Many materials in the car are light, but very strong.

The nose cone of the vehicle is made of strong, light fibreglass.

The spoiler and the wings are made from two materials. The inner core is light. It's made of polystyrene. The outer skin is hard and made of fibreglass.

The frame is light, but very tough and rigid. It's made of cromoly, a steel alloy.

The radiator is made of aluminium. The aluminium is coated with ceramic. These two materials are corrosion-resistant.

The engine and pistons are made of a light aluminium alloy. Each piston inside the engine is coated with a heat-resistant ceramic.

The wheels are made of a strong, light aluminium alloy. The tyres are made of a tough rubber composite.

an *alloy* is a mixture of two or more metals
a *composite* is a mixture of two types of material
fibreglass is a composite. It is a mixture of a plastic and a ceramic

BrE *tyre*; AmE *tire*

Part	What's it made of?	What are its properties?
nose cone	(1)	(2)
spoiler and wings	coated with (3)	(4)
wheels	(5) alloy	(6)
tyres	(7) composite	(8)
pistons	coated with (9)	(10)
frame	(11)	(12)
radiator	(13)	(14)

Materials | 6 | 45

3 Buying

Listening 1 ▶🎵35 Listen and complete the customer call form.

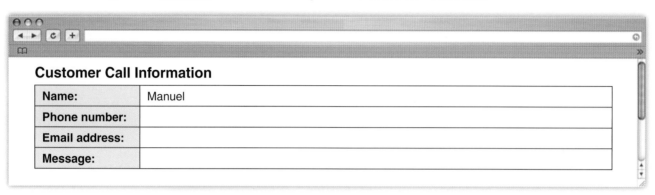

Customer Call Information	
Name:	Manuel
Phone number:	
Email address:	
Message:	

2 ▶🎵36 Listen and write the correct email and web addresses.

When you hear this	write this
1 waleed at sports dot com	waleed@sports.com
2 adam at city dot co dot U, K	
3 theo walcott, that's T-H-E-O then W-A-L-C-O-T-T at goalfeast, that's G-O-A-L-F-E-A-S-T all one word dot com	
4 C dot ronaldo, that's R-O-N-A-L-D-O at back-of-the-net, that's B-A-C-K dash O-F dash T-H-E dot net	
5 www dot toyota, that's T-O-Y-O-T-A dot com forward slash customer dash support	
6 www dot orascom, that's O-R-A-S-C-O-M dot com dot E-G forward slash sales underscore one	

Speaking 3 Work in pairs. Dictate the addresses to your partner.

Student A. Turn to page 114. Student B. Turn to page 118.

Listening 4 ▶🎵37 Listen to this phone conversation and complete the questions.

- *What's your surname, please?*
- *It's Lint.*
- *Could you (1) _____ that, please?*
- *Lint.*
- *Could you (2) _____ that, please?*
- *L-I-N-T.*
- *(3) _____ T or D?*
- *It's T. T for teacher.*
- *Thanks. And what's the product number?*
- *It's 17-305.*
- *(4) _____ 17 or 70?*
- *Teen. Seven<u>teen</u>. One seven.*
- *Right. Thanks.*

> Never put a stress on the *-ty* in numbers like *thirty*, *forty*, *fifty* and so on.
> Tip: say *seventy* but *seven<u>teen</u>* to make the difference clear.

Speaking 5 Practise the phone call in pairs. Then change roles.

Task 6 Work in pairs. Buy sports equipment over the telephone.

Student A. Turn to page 116.

Student B:
1 You are the customer. Circle three items you would like to buy, and circle the features you want (size, colour, material), and the price. Then phone up the shop and place your order. You can either make up details (e.g. names, phone numbers, etc.) or use your own.
2 Then change roles. You are now the sales person in the sports shop. Ask Student A questions and complete this order form.

no. = number
# = number

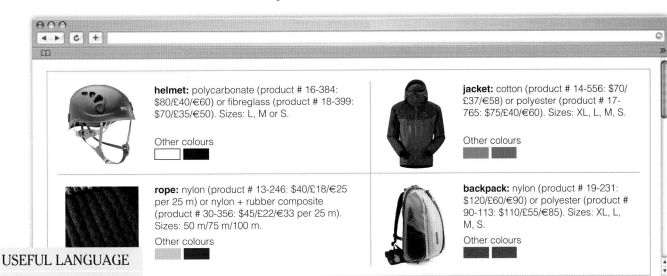

helmet: polycarbonate (product # 16-384: $80/£40/€60) or fibreglass (product # 18-399: $70/£35/€50). Sizes: L, M or S.

Other colours

jacket: cotton (product # 14-556: $70/£37/€58) or polyester (product # 17-765: $75/£40/€60). Sizes: XL, L, M, S.

Other colours

rope: nylon (product # 13-246: $40/£18/€25 per 25 m) or nylon + rubber composite (product # 30-356: $45/£22/€33 per 25 m). Sizes: 50 m/75 m/100 m.

Other colours

backpack: nylon (product # 19-231: $120/£60/€90) or polyester (product # 90-113: $110/£55/€85). Sizes: XL, L, M, S.

Other colours

USEFUL LANGUAGE

- What's your name/phone number/email address?
- Could you spell/repeat that, please?
  Is that six<u>teen</u> or <u>six</u>ty?
- What's the product name/number?
- What colour/size/material would you like/do you need?
- Do you want to pay in dollars ($), sterling (£) or euros (€)?
- How many would you like/do you need?

Name	
Phone no.	
Email address	
Order	

Product name	Product no.	Colour	Size	Material	Price	Quantity

Social English 7 **🔊 38** Listen to three telephone calls. Mike (M) is phoning his friend John (J).

	1	2	3
J	Hello?	Hello?	Hello. John Davis here.
M	*Hello. Is that John?*	*Hello. Is that John?*	*Oh hi, John. This is Mike.*
J	Yes?	Yes. Is that Mike?	Hi, Mike.
M	*It's Mike.*	*Yes, it's me. Hi. How are you?*	*Hi. How are things?*
J	Oh hi, Mike.	Fine, thanks. How about you?	Great, thanks. How are you?
M	*Hi. How are you?*	*I'm fine. (Begin your call).*	*Good. (Begin your call).*
J	OK, thanks. How are you?		
M	*Fine. (Begin your call).*		

8 Work in pairs. Practise short phone calls, using your own names.

Review Unit C

1 Look at the pictures and give instructions with the words in the box.

| bend | close | cut | drive in | grip | loosen | measure | open |
| pull out | put | put on | strike | take | take off | tighten | use |

Examples: 1 Grip the nail. Use a pair of pliers. 2 Pull out the nail.

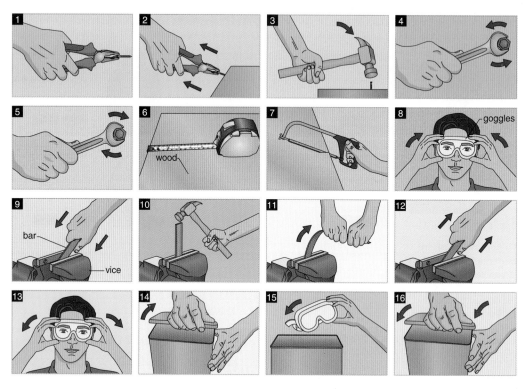

2 Say what is happening in the pictures in 1.

Example: 1 He's gripping the nail. He's using a pair of pliers.

3 Correct the mistakes in these sentences.

1. Water boils at 32°F. (freeze)
   <u>Water doesn't boil at 32°F. It freezes.</u>

2. Hot water sinks to the bottom of a tank. (rise / top)
   _____

3. Cool air rises to the top of a room. (sink)
   _____

4. Hot air sinks to the bottom of a room. (stay / top)
   _____

5. The Sun's rays cool the water in the solar panel. (heat)
   _____

4 Complete the dialogue with the correct form of the verbs in brackets.

- How does the thermosiphon (1) _____ (work)?
- Well, the cold water (2) _____ (enter) *the system through the inlet. The water pressure* (3) _____ (push) *the water around the system.*
- So how (4) _____ (do) the water (5) _____ (become) hot?
- It (6) _____ (flow) *into the panel and the sun's rays* (7) _____ (heat) *it. The warm water* (8) _____ (rise) *to the top of the panel and it* (9) _____ (pass) *from the panel into the tank.*
- (10) _____ (do) the tank (11) _____ (have) a heater?
- No, it (12) _____ (do not). *The hot water* (13) _____ (stay) *at the top of the tank. If you* (14) _____ (open) *the valve, the hot water* (15) _____ (flow) *from the top of the tank to the outlet.*

5 Identify the equipment from the description.

| cable   fan   pump   radiator   solar panel   thermostat |

1. It converts energy from the Sun into heat or electricity.
2. It pushes water around a water supply system, or around a car engine.
3. It blows cold air onto a car radiator and cools the water inside it.
4. It controls the temperature of water or air in a heating or cooling system.

6 There's a problem with the forklift truck. Say what's going wrong.

1. I <u>'m pressing</u> (press) the accelerator pedal, but the truck <u>isn't going</u> (not go) faster.
2. He _____ (pull) the lever back, but the forks _____ (not rise).
3. You _____ (push down) the brake pedal, but the truck _____ (not slow).
4. I _____ (slide) the lever forwards, but the forks _____ (not tilt) upwards.
5. He _____ (pull) the direction lever backwards, but the truck _____ (not reverse).
6. You _____ (move) the direction lever forwards, but the truck _____ (not go) forwards.

Review Unit C

7 Complete the sentences with *bend* or *break* and other words.

1. Polyester is a tough material. You can't _____ it easily.
2. Concrete is a rigid material. It doesn't _____ easily.
3. Polycarbonate is a hard material. It _____.
4. This glass is brittle. You _____.
5. These plastic rulers are very flexible. They _____.

8 Draw a line from each word to its opposite.

rise enter into heavy strong
go in inlet push light pull
sink open out of tough hard
go out to outlet soft
leave weak brittle flexible
close go down from rigid go up

9 Complete the sentences with the correct form of the verb in the box.

| boil freeze melt rise sink stretch |

1. If you heat water to 100°C, it _____.
2. If you cool water to 0°C, it _____.
3. If a heater warms the air in a room, the air _____.
4. If an air conditioner cools the air in a room, the air _____.
5. If you heat steel bars to 1400°C, they _____.
6. If you pull a copper wire very hard, it _____.

10 Identify the material from the description. Choose from the words in the box.

| aluminium ceramic polycarbonate polystyrene rubber steel |

1. Sunglasses are made of this material. It's a hard and tough plastic.
2. You can stretch this material and you can bend it, but it doesn't break.
3. You can heat this material to a high temperature, but it doesn't burn or melt. They use it in spark plugs.
4. Parts of aeroplanes are made of this material. It's a strong, light, corrosion-resistant metal.

Review Unit C

11 Make dialogues about the parts of a racing car.

1 nose cone / fibreglass / strong and light
2 pistons / aluminium alloy / light and corrosion-resistant
3 frame / cromoly / tough and rigid
4 tyres / rubber composite / tough
5 radiator / aluminium and ceramic / corrosion-resistant
6 outer skin of spoiler / fibreglass / hard

A: *What's/What are the … made of?*

B: *It's/They're made of … .*

A: *Why do they/we/you use … ?*

B: *Because it's … .*

12 Complete the text with the correct form of the verbs in brackets.

This is how you test the properties of the material. You put the material into the multi-test machine. Then the machine does four tests on it. In the first test, a hammer (1) _____ (strike) the material with a 50 kg weight. In the second test, two pairs of jaws (2) _____ (pull) the material with a weight of 80 kg. In the third test, a heavy weight of 100 kg (3) _____ (press) the material down. In the fourth test, two sharp knives (4) _____ (scratch) the material with weights of 10 and 20 kg.

OK, now I'm demonstrating the four tests in action. Watch carefully. Here's the first test. The hammer (5) _____ (strike) the bar. Can you see? The bar isn't breaking. Here's the second test. It's starting now. The jaws (6) _____ (pull) the material. Can you see? The material (7) _____ (not stretch). Now the third test is taking place. The heavy weight (8) _____ (press) the material down. Can you see that? The material (9) _____ (not break). And now here's the fourth and final test. The knives (10) _____ (scratch) the material.

Projects

13 Find out what these word parts mean. Then find other words with the same word part.

Word part	Meaning of word part	Example of word	Meaning of word
sol-		1 *solar* 2	1 2
poly-		1 *polytechnic* 2	1 2

14 Find out about materials you use in your industry. Make your own table and complete it.

Example:

Industry: Aerospace		
Application	Material	Property
Wing parts	Aluminium alloys	Light, strong, corrosion-resistant

Review Unit C

7 Specifications

1 Dimensions

Start here 1 What do you know about this bridge?

1 What's it called?
2 Where is it?
3 How high is it?

Listening 2 ▶️ 39 Listen to part of a TV programme about the bridge. Check your answers to 1.

3 Work in pairs. Which of the following can you see in the photo?

cable   deck   pier   pylon   span

4 ▶️ 40 Listen to the next part of the TV programme and complete the specifications of the bridge.

BrE: *metre, millimetre, centimetre.*
AmE: *meter, millimeter, centimeter.*

Don't add -s to abbreviations of units.
say: *one hundred metres / kilometres*; write: *100 m / 100 km*

Millau Bridge: specifications				
Structure	(1) *cable-stayed*	Length of outer spans	(7)	m
Completion date	(2) *December 2004*	Number of piers	(8)	
Material: cables and deck	(3)	Height of pylons above deck	(9)	m
Material: piers	(4)	Height of deck above water	(10)	m
Total number of spans	(5)	Length of deck	(11)	km
Length of inner spans	(6) m	Width of deck	(12)	m

Vocabulary

5 Complete the table.

Adjective	high	long	_____	wide
Noun	_____	_____	depth	_____

6 Complete the sentences with the correct word in brackets.

1 The _____ of the road is 6 m. (wide/width)
2 The river is 230 km _____. (long/length)
3 The sea has a _____ of 330 m. (deep/depth)
4 These pylons are over 80 m _____. (high/height)
5 These oil wells are more than 700 m _____. (deep/depth)
6 The total _____ of the road is about 120 km. (long/length)
7 The tunnel is 15 m _____. (wide/width)
8 The _____ of the bridge is 130 m. (high/height)

Language

| How | high
wide
long
deep | is it?
are they? | It's
They're | 2
10
100
1000 | millimetres
centimetres
metres
kilometres | high.
wide.
long.
deep. |

Speaking

7 Make questions about the Millau Bridge. Use the specification chart in 4.

8 Work in pairs. Ask and answer your questions in 7.

Example:
TV presenter: How long are the inner spans?
Engineer: They're 342 metres long.

Task

9 Work in pairs. Find out the specifications of your partner's bridge.

Student B. Turn to page 118.

Student A:
1 Ask Student B questions about the Akashi-Kaikyo Bridge. Complete your specifications chart.
2 Then change roles. Turn to page 114 and answer Student B's questions about the Rion-Antirion Bridge.

Akashi-Kaikyo Bridge: specifications	
Type of structure	Suspension
Country	
Piers (number)	
Span (length)	
Deck (above water)	
Deck (length)	
Water (max depth)	
Water at main pier (depth)	

The Akashi-Kaikyo Bridge

2 Quantities

Start here 1 Try the quiz. Match the names of the buildings to the pictures. Write the number and the approximate height of each building.

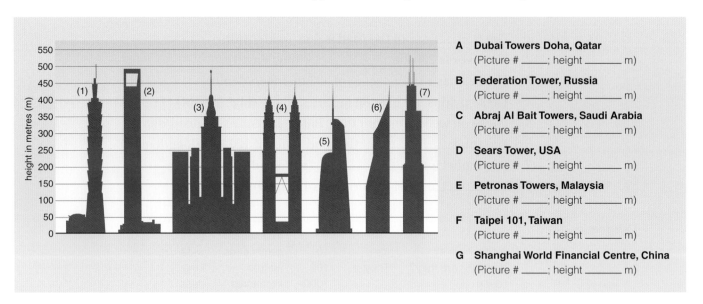

A Dubai Towers Doha, Qatar
(Picture # _____; height _____ m)

B Federation Tower, Russia
(Picture # _____; height _____ m)

C Abraj Al Bait Towers, Saudi Arabia
(Picture # _____; height _____ m)

D Sears Tower, USA
(Picture # _____; height _____ m)

E Petronas Towers, Malaysia
(Picture # _____; height _____ m)

F Taipei 101, Taiwan
(Picture # _____; height _____ m)

G Shanghai World Financial Centre, China
(Picture # _____; height _____ m)

2 ▶ 🔊 41 Listen and check your answers to 1.

Reading 3 Read the FAQs from the website and match them to the answers.

BrE *lift* = AmE *elevator*

write: *8000 m²*; say: *eight thousand square metres*.
write: *250,000 m³*; say: *two hundred and fifty thousand cubic metres*.
write: *5 kg*; say: *five kilograms* or *five kilos*.

This is Taipei 101. It is currently the highest in the world. Here are some frequently asked questions (FAQs) about the building.

1 *How high is Taipei 101?*
2 *What's the footprint of the building?*
3 *How many storeys does it have?*
4 *How do you get to the top?*
5 *What's the building made of?*
6 *How much steel and concrete is in the building exactly?*

A About 700,000 tonnes.
B By super-fast elevator. The building has two high-speed elevators. Each elevator travels at 17 m/s.
C 101.
D It towers above Taipei at the amazing height of over 508 metres.
E Reinforced concrete, steel, aluminium and glass.
F The base of the building has an area of about 450 m².

54   7 Specifications

Language Countable nouns can be both singular and plural. Examples: *screw, nail, bottle*.
Uncountable nouns are always singular. Examples: *concrete, cement, sand, oil*.

screws are countable			**cement** is uncountable	
a / one	screw		some	cement
some / two	screw	-s		
a bag of / two bags of			a bag of / two bags of	

| Do you need | some/any | screws? / cement? | How | many (screws) / much (cement) | do you need? |

4 Complete the dialogue with the words in the box.

any   how   many   much   some   What colour   What size

- Good morning. Can I help you?
- Hello. Do you have (1) _____ screws?
- Certainly. (2) _____ do you need?
- Ten mil.
- OK. And (3) _____ _____ do you need?
- Fifty, please.
- Right. So that's fifty 10 mil screws. Anything else?
- Yes. I need to buy (4) _____ paint, please.
- (5) _____?
- Black.
- OK. So (6) _____ _____ black paint do you need?
- Six large tins, please.
- Anything else?
- No, that's all, thanks.

5 Make similar dialogues with your partner. Use the questions in the box and the information from the table.

How many?   How much?
What colour?   What kind?
What size?   What type?

To buy ...		
Item	Quantity	Kind, size or colour
screws	50	10 mm
paint	6 large tins	black
glue	2 tubes	superglue
nuts	30	15 mm
oil	15 L	motor oil
bolts	60	25 mm
cement	20 bags	white
nails	2 packets of 50	20 mm

write: *15 L*; say: *15 litres*

 tin    tube    bag    packet

Specifications 7

3 Future projects

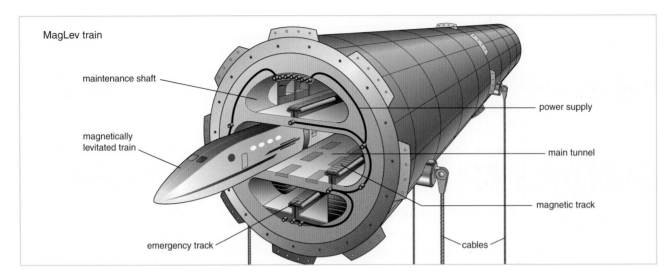

MagLev train — maintenance shaft, magnetically levitated train, emergency track, power supply, main tunnel, magnetic track, cables

Start here

1. Work in pairs. Look at the picture. What is it? How does the vehicle move?
2. 🔊 42 Listen to this radio interview and complete the specification chart.

Trans-Atlantic MagLev Tube	
Location of tube	(1) *Under the Atlantic Ocean from Britain to the USA*
Possible date of completion	(2) *2100*
Length	(3) km
Depth below sea level	(4) m
Number of cables	(5)
Speed of train	(6) km/h
Source of power for train	(7)

Language

Use *will* and *won't* to predict a future fact or event.

They/We My company The engineers	will 'll will not won't	build it in 2050.	
When	will	they/you	build it?	In 2050.
	Will		build it in 2050?	Yes, they will. / No, they won't.

3. Disagree with each statement.

   1. The engineers will start the tube in 2020. (2080)
   2. The tube will be under the Pacific Ocean. (Atlantic)
   3. The tube will connect Britain with Europe. (the USA)
   4. The train will use diesel. (magnetism)
   5. The tube will contain compressed air. (a vacuum)
   6. The trains will travel at 11,000 km/h. (8000 km/h)

   Example: 1 They won't start the tube in 2020. They'll start it in 2080.

7 | Specifications

Reading **4** Read this interview and produce a specifications chart for the bridge (see 2 on page 56). Use the words in the box.

completion date   deck   height   length   materials   pier   pylon   span

Bridge of the Future:
Europe-Africa Bridge

RadioTech presenter Tom Burns interviews engineer Galal Hamdy.

Tom: What project are you working on now?
Galal: We're designing the world's longest bridge.
Tom: Where will it be?
Galal: Between Morocco and Spain. It'll connect Europe with Africa.
Tom: What are the specifications of the bridge?
Galal: It will be almost 15 km long. In our design, the bridge will have two spans. Each span will be 4800 m long.
Tom: That's a very long span. How will that be possible?
Galal: The bridge will have three steel pylons, on concrete piers. The pylons will be 1000 m high. The deck will be very light and strong. It'll be made of fibreglass.
Tom: Many engineers think you won't be able to build this bridge.
Galal: I don't agree. I think we'll complete it around 2030.

Speaking **5** Work in pairs. Ask and answer questions about the specifications of the bridge.

A: *How long will the bridge be?*   B: *It will be almost 15 km long.*

6 Here is a possible project schedule for the Europe-Africa Bridge. Roleplay an interview between a TV presenter and an engineer.

Task	2024	2025	2026	2027	2028	2029	2030	2031	2032
1 lay foundations									
2 build piers									
3 put pylons on piers									
4 attach cables to pylons									
5 make deck									
6 fix deck to cables									
7 build roads									
8 open bridge									

TV Presenter: When will you build the piers?
Engineer: We'll start in 2026 and finish in 2027.

Social English **7** How do you think the world will change in the next 20 years. Think about technology, social, political and legal changes.

Example: Computers will control more things in our homes.

8 Reporting

1 Recent incidents

Start here 1 Work in pairs. Look at the photo and say what's happening. List five common problems you can have with a car.

2 ▶ 43 Listen to this phone call and complete the details in the form.

Crash Recovery Co Ltd Online customer call information. Enter details

Customer name	(1)
Car Registration No	(2)
Location: Road	(3)
Between Junction (4)	and Junction (5) Going (6)
Problem:	(7) *The exhaust pipe*

Listening 3 ▶ 44 Listen to the phone calls and match them with the pictures.

4 Complete the sentences with the verbs in the box. Put two words in each gap.

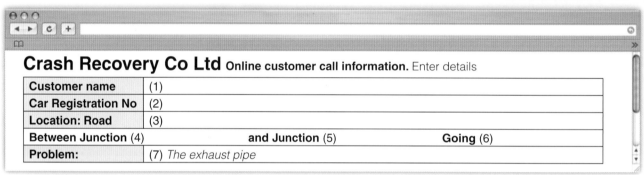

broken cut driven fallen had happened have/has lost taken

1 Is that Security? Thieves *have broken* into my office. They _____ my computer.
2 Is that the IT hotline? Something _____ to my computer. I _____ all my data.
3 I need an ambulance, quickly. My daughter _____ downstairs. She _____ her leg.
4 Is that Crash Recovery? I _____ an accident. I _____ my car into a bridge.

Language You form the *present perfect* with *have/has* + past participle.
- You can use the present perfect to talk about *recent* actions: *My car has broken down. I've changed the tyre.*
- The present perfect does not go with dates, times or time expressions such as *yesterday, a week ago, in 2005*. Use the past simple with these expressions.

5 Check you know the past participle of these verbs. Which ones are *irregular*?

buy check crash fall order put repair
sell send speak steal take write

58 8 | Reporting

Speaking

6 Work in pairs. Make short dialogues.

A is the supervisor in a car repair workshop. *B* is a mechanic in the workshop.
1 check the brakes ✓ repair the tyres ✗
2 order those new parts ✓ buy those tools ✗
3 change the tyres ✓ clean the spark plugs ✗
4 phone the customer ✗ speak to our supplier ✓
5 write that report ✓ send that email ✗
6 put in the new fuses ✗ take out the old lamps ✓

A: *Have you checked the brakes?*
B: *Yes, I have.*
A: *Good. What about the tyres? Have you repaired them?*
B: *No, I haven't. I'll do it now.*

7 Try this memory test.

- Look at the picture on page 117 for one minute.
- Then look at the picture below. How many differences are there? Compare with a partner.

8 It is now 10.16 am. Explain what has happened in the picture since 10:12 am. Use the nouns and verbs in the box.

| beam bricks bucket builder crane digger |
| hard hat jacket scaffolding sledgehammer |

| climb down drive fall over lower move back pick up put put on raise take off |

Example: 1 Two builders have taken off their jackets.

2 Damage and loss

Start here 1 Do you have any damaged tools or equipment? Describe the damage to your partner.

Vocabulary 2 Do you remember the verbs in the box? Match them with the pictures.

bend   break   burn   crack   cut   dent   scratch   tear

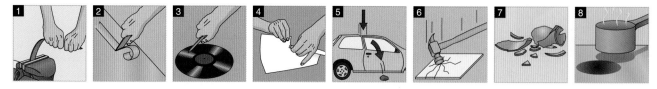

Task 3 Correct the mistakes in this checklist.

Quick Start guide

Check all these items are in the box and in good condition.
If any items are damaged or missing contact Customer Services immediately.

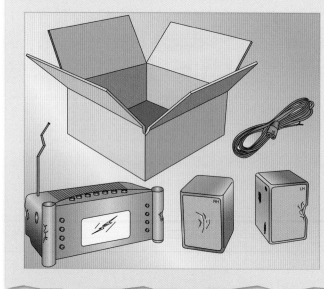

item	in box	condition
radio	✓	damaged
• radio antenna		OK
• body of radio		cracked
• display screen		OK
power cable with plug	no plug	cable OK
4 AA batteries	✓	OK
1 user manual	no manual	–
1 pair headphones	✓	OK
1 LH external speaker	✓	OK
1 RH external speaker	✓	OK
2 cables for speakers	✓	OK

Listening 4 ▶ 45 Look at the picture in 3. Listen to the telephone conversation and check the list.

Speaking 5 Look at the picture in 3 again. Make sentences about the damage and the things that are missing. Use these sentence patterns.

Ways to report damage	Ways to report something missing
The screen is scratched.	The manual is missing.
There's a scratch on the screen.	There's no manual in the box.
The speakers are dented.	The cable has no plug. / The cable doesn't have a plug.
There are some dents on the speakers.	There's no plug on the cable.

8 Reporting

Language

Focus on action			Focus on result of action		
I have	dented	the radio.	The radio	is	dented.
He has	broken	the speakers.	The speakers	are	broken.
	past participle				adjective

6 Rewrite the sentences in the same way as in the table above.

Focus on action	Focus on result of action
1 I've scratched the display screen.	
2 Someone has bent the antenna.	
3 I've burnt the body of the radio.	
4 Someone has dented the top of the speaker.	
5 They've cracked the cover of the plug.	
6 Someone has torn the user manual.	

7 Complete the sentences with the correct form of the words in the box.

bend   crack   cut   dent   scratch   tear

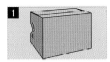    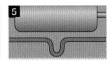

1   The side of the box is _____.
2   The lenses of the goggles are _____.
3   The surface of the road is _____.
4   The insulation of the cable is _____.
5   The pipe below the tank is _____.
6   The overalls are _____.

8 Rewrite the sentences in 7 to give the same meaning.

There's a … . / There are some … .

Example: 1 There's a dent in the side of the box.

Task

9 Work in pairs. Find out the damage to your partner's car.

Student A:
1   Ask Student B questions about the damage to their car. Label your diagram.
2   Then change roles. Turn to page 115.

Student B. Turn to page 116.
● What's the problem?
○ *The door is scratched.*
● Which door?
○ *The back / front nearside one.*
● Anything else?

front ≠ rear
The steering wheel is always *offside*.

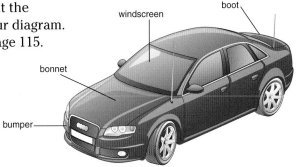

3 Past events

Start here 1 Work in pairs. When did these events happen?

Give the approximate year of the first ...
1. space station
2. telescope in space
3. man in space
4. space tourist
5. satellite
6. spacewalk
7. man on the Moon
8. shuttle in space
9. crew to enter the International Space Station
10. European navigation satellite

Reading 2 Read this chart and check your answers to 1.

Key dates in the history of space flight

Event	Date
1 The Russians launched Sputnik, the first satellite.	5 October 1957
2 Yuri Gagarin became the first man in space.	12 April 1961
3 Leonov made the first walk in space.	18 March 1965
4 The first men, Armstrong and Aldrin, landed on the Moon.	20 July 1969
5 The Russians launched the first space station, Salyut 1.	19 April 1971
6 The Americans put the first shuttle into space.	12 April 1981
7 NASA sent the Hubble telescope into space.	24 April 1990
8 The first crew entered the International Space Station.	2 November 2000
9 The first space tourist flew into space.	28 April 2001
10 The Europeans launched Galileo, a global navigation satellite.	28 December 2005

Language This is the *past simple* form of the verb.

- You can use it to talk about *past events*.
- Use the past simple with dates, times or expressions such as: *yesterday, last year, When?*

When	did	he/she/it/they/we/you	go travel	there?	
		He/She/It/They/We/You	went travelled	there	in 2007.

Speaking 3 Make questions and answers about the table in 2.

A: *When did the Russians launch Sputnik?*
B: *They launched it on the 5th of October 1957.*
(or: *They launched it in 1957.*)

Use *on* for the exact day: on the 14th of May 2005.
Use *in* for a month or a year: in May; in 2005.

Vocabulary *ago = before now*

You can say *the fifteenth of November* or *November the fifteenth*.

If it is the 15th of November today …	If it is 10.15 now …
• two days ago = 13th November	• five minutes ago = 10.10
• two weeks ago = 1st November	• an hour ago = 9.15
• two months ago = 15th September	• two hours ago = 8.15

4 Write the name of this month on the calendar. Put a circle round today's date. Then say what the following dates are.

1 today
2 yesterday
3 the day before yesterday
4 two days ago
5 one week ago
6 two weeks ago

			1	2	3	4	5
6	7	8	9	10	11	12
13	14	15	16	17	18	19
20	21	22	23	24	25	26
27	28	29	30	31		

more than 50 ≠ less than 50

5 Make statements about the chart in 2 using **ago** and approximate years from today's date.

Example: 1 The Russians launched Sputnik more than 50 years ago.

6 Listen and complete the phone call.

- Hello, Electronic Repairs. Don speaking. How can I help you?
- Hi. My name's Ben Jones. I've (1) _____ my MP3 player. Can you repair it?
- OK, sir. What's the model number?
- It's a Super 30 GB.
- And when did you (2) _____ it?
- Er, let's see … Yes, I (3) _____ it on the 18th of August.
- And what's the problem?
- I've (4) _____ it and I've (5) _____ the screen.
- And, er … when did you (6) _____ the screen?
- Yesterday.
- OK, bring it into the shop and I'll look at it.
- Thanks. Bye.

7 Work in pairs. Make similar phone calls.

	Item 1	Item 2	Item 3
Item:	MP3 player	mobile phone	laptop
Model no:	60 GB	9300	Travel 380
Date of purchase:	15th February	13th October	21st July
Damage:	dented cover	dropped in water	broken cover
Date of damage:	three days ago	day before yesterday	two weeks ago

Social English

8 Make a list of interesting things you have done in your life, with their dates.

- *climbed Mont Blanc in June 2006*
- *snorkelled in the Red Sea in August 2007*

9 Tell other students in your class about your list.

Review Unit D

1 Make questions for these answers.

1. It's about 50 m wide. (the road)
   How wide is the road?

2. They're 90 m high. (the pylons)
   _____

3. It's more than 2 km long. (the deck of the bridge)
   _____

4. It's about 35 m in height. (the scaffolding)
   _____

5. They're 15 m deep. (the foundations of the building)
   _____

6. They're about 12 m in length. (the steel beams)
   _____

2 Change these nouns to adjectives

1. depth   _____
2. height   _____
3. width   _____
4. length   _____

3 Rewrite the sentences to give the same meaning.

1. What is the height of the bridge?
   How high is the bridge?

2. The height of the tower is 46 m.
   The tower is _____.

3. What is the depth of the sea under the bridge?
   How _____?

4. The length of the new road is 355 km.
   This new road is _____.

5. What are the widths of the screws?
   How _____?

6. The depth of the well is more than 30 m.
   The well is _____.

4 Make questions for these answers.

1. It has ten. (storeys / building)
   How many storeys does the building have?

2. He needs 20 kilos. (cement / builder)
   _____

3. They're using two. (cranes / men)
   _____

4. It needs about 4 litres. (oil / car)
   _____

5. I'm buying 150. (screws / you)
   _____

6. They can carry about 50 cubic metres. (concrete / ten trucks)
   _____

5 Read the text. Label the diagram with all the parts and dimensions in italics.

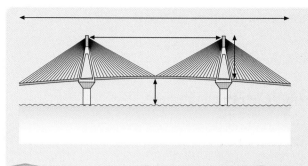

This cable-stay bridge has 20 *cables*. The *deck* of the bridge is *1.2 km* long, and is *185 m* above water level. Each *pier* is *35 m* wide. The *span* between the two piers is *832 m* long. Each *pylon* is *45 m* high above the road deck.

6 Work in pairs. Order what you need to build the Burj Dubai skyscraper.

trucks / 40,000	cranes / 3	steel poles / 12,000
concrete / 150,000 m³	steel / 25,000 tonnes	aluminium / 15,000 tonnes

A: *I need to order some concrete/some trucks.*
B: *OK. How much concrete/How many trucks do you need?*
A: *I need … .*

7 Complete the dialogue.

- ● Engineers are planning to build a tunnel under the sea.
- ○ *Where will the tunnel be?*
- ● *It'll be between Spain and Morocco.*
- ○ How long (1) _____ be?
- ● It (2) _____ .
- ○ How many (3) _____ have?
- ● It (4) _____ .
- ○ How (5) _____ ?
- ● It (6) _____ .
- ○ How (7) _____ ?
- ● It (8) _____ .
- ○ When (9) _____ the engineers _____ ?
- ● They (10) _____ .

Location:	• Between Spain and Morocco
Length:	• 40 km
Number of railway lines:	• 2
Width:	• 8 m
Depth (below sea level):	• 300 m
Completion date:	• 2025

8 Answer these questions.

1 Did they complete the Millau Bridge in 2000? (2004)
   No, they didn't. They completed it in 2004.

2 Have you ever worked in an electronics company? (video shop)
   _____

3 Will they build a bridge from Africa to Europe? (a tunnel)
   _____

4 Are they constructing the tunnel now? (planning and designing)
   _____

5 Has NASA ever put men on Mars? (the Moon)
   _____

6 Did Russia launch the first satellite in 1960? (1957)
   _____

9 Rewrite the sentences using the present perfect tense.

Remember: don't use a time expression (such as *yesterday* or *an hour ago*) with the present perfect.

1 My car broke down five minutes ago.
   My car has broken down.

2 NASA launched the space shuttle fifteen minutes ago.
   _____

3 A virus attacked our office computers two hours ago.
   _____

4 I wrote the email and I sent it to the customer yesterday.
   _____

5 The technician took the hard drive out of the computer an hour ago.
   _____

6 The exhaust pipe fell off my car ten minutes ago.
   _____

10 Look at the pictures. Say what's missing, in three different ways.

Example: 1 The wheel has no wheel nuts. / The wheel doesn't have any wheel nuts. / There are no wheel nuts on the wheel.

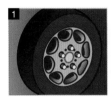 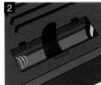 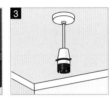 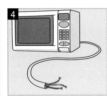 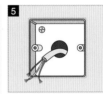 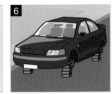

11 Complete the table.

Focus on action	Focus on result of action
1 He's dented the front bumper.	The front bumper is dented.
2 You've broken the windscreen.	
3 Someone has burnt the rear seat of the car.	
4 We've bent the poles of the scaffolding.	
5 They've torn the safety jackets.	
6 Someone has scratched the rear panel of the car.	

12 Complete the table.

1 He's *bent* the antenna.	The antenna is _____.	There's a small _____ in the antenna.
2 The fire has *burnt* the walls.	The walls are _____.	There are two large _____ on the walls.
3 You've *cracked* the window.	The window is _____.	There are some _____ in the window.
4 I've *torn* my shirt.	My shirt is _____.	There's a _____ in my shirt.

13 Rewrite these sentences to give the same or similar meaning.

1. There's a scratch on this cover.   This cover is _____.
2. There are no wheels on the car.   The car has _____.
3. The cables don't have any plugs.   There are _____.
4. The windscreens are cracked.   There are some _____.
5. There's no workshop manual in this garage.   This garage doesn't _____.
6. There is a dent in the roof of the car.   The roof _____.

14 Complete this dialogue with the correct form of the verb in brackets.

● *Where did you buy your safety equipment?*
○ I (1) _____ (buy) it online, over the Internet.
● *That's good. How did you (2) _____ (find) the website?*
○ I (3) _____ (find) it through Google. I (4) _____ (key) in the words 'safety gear'.
● *How (5) _____ (you / pay) for it? Did you (6) _____ (use) your own bank card?*
○ No, no. My company (7) _____ (give) me a credit card last week. I (8) _____ (use) that.
● *That's great. When (9) _____ (you / receive) the goods?*
○ They (10) _____ (come) yesterday, by express mail.

15 Write a description of this water tower and how it works. Use the notes below.

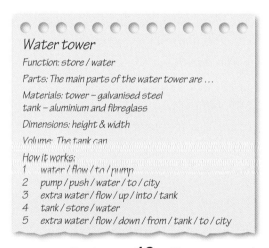

Water tower
Function: store / water
Parts: The main parts of the water tower are …
Materials: tower – galvanised steel
tank – aluminium and fibreglass
Dimensions: height & width
Volume: The tank can
How it works:
1 water / flow / to / pump
2 pump / push / water / to / city
3 extra water / flow / up / into / tank
4 tank / store / water
5 extra water / flow / down / from / tank / to / city

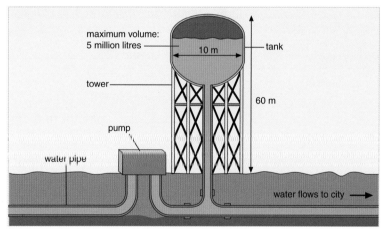

Projects   **16** Choose one of these projects and follow the instructions.

1. Find out some facts about a famous structure (for example a bridge or building). Write a short article about it for an in-flight tourist magazine.

2. Design a new bridge, tunnel, or transport link (e.g. railway line or hovercraft route) to connect two places. Find out some facts about the location (for example, the width of a lake, the depth of the lake, the height of the land beside the lake, and so on). Write a short article about it for a technical magazine.

   a) Draw a simple diagram of your design. Mark the dimensions.
   b) Produce a specifications chart.
   c) Write a short description.

9 Troubleshooting

1 Operation

Start here

1 Work in pairs. How does this vehicle move? Discuss with your partner.

2 What do the main parts do? Complete the chart.

Part	Function
	drive the fan
	pull the air in + force the air down
	control the speed and acceleration
	steer the airboard
	support the rider

Listening

3 🔊 47 Listen and check your answers.

4 Listen again and complete the dialogue.

- Look at the airboard. You can see the five main parts: the body, the engine, the fan, the handlebar and the two levers. The body (1) <u>supports</u> the rider and the engine (2) _____ the fan. The handlebar (3) _____ the airboard left and right.
- Ah yes, I see. So what (4) _____ the fan (5) _____?
- It (6) _____ the air in and (7) _____ it downwards.
- Right. And what (8) _____ the two levers (9) _____?
- They (10) _____ the speed and acceleration of the airboard.

downwards ≠ upwards

Language

What	does	the engine	do?	It	drive	-s	the fan.
	do	the lever	-s		They	control		the speed.

5 Make short dialogues about the parts of the airboard.

1 fan / cool the engine? no ➔ push air downwards
2 engine / drive the wheels? no ➔ drive the fan
3 levers / stop the airboard? no ➔ increase the speed
4 handlebars / control the brakes? no ➔ steer the airboard

A: *Does the fan cool the engine?*
B: *No, it doesn't.*
A: *So, what does it do?*
B: *It pushes air downwards.*

Reading   **6**   Read this article from a technical magazine and answer the questions below.

THE AIRBOARD how it works

You stand on the airboard and ride it like a skateboard. The board moves on a cushion of air, like a small hovercraft. It has a fibreglass body, an engine, a large fan,
5 a flexible rubber skirt, a friction wheel, a handlebar and two levers.

The engine and the fan are mounted on the body. The skirt and the friction wheel are suspended from the body. The handlebar
10 is mounted on the body, at the front. The levers are attached to the handlebar.

The engine drives the fan. The function of the fan is to suck air in and to force it downwards. This pushes the vehicle
15 upwards and propels it forwards. On the body there is a fibreglass platform. This supports the rider. The skirt contains the air and the cushion of air supports the airboard. The rider uses the handlebar to
20 steer the board. One lever controls the speed of the engine and the fan. The other lever controls the friction wheel. The friction wheel touches the ground for one or two seconds and accelerates the airboard into
25 the air. If you want to stop, simply release the levers.

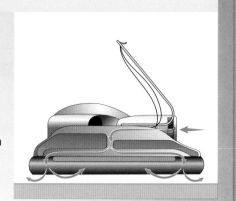

press ≠ release

1 What is the friction wheel for?
2 Is the skirt above or below the body? What is it made of? Can you bend it?
3 Which part of the airboard does the rider stand on?
4 What happens if you take your hands off the levers?
5 Does *propel* (line 15) mean *pull*, *push*, *hold* or *control*?
6 Find words which mean the opposite of (1) *backwards* (2) *upwards*.

Language   **7**   Rewrite the sentences to give the same meaning.

1 The purpose of the handlebar is to steer the airboard.
2 The job of those levers is to control the speed of the airboard.
3 The function of the friction wheel is to accelerate the airboard.
4 The purpose of the fan and the engine is to propel the airboard forwards.
5 The function of the skirt is to hold the air and to support the airboard.
6 The job of the body and the platform is to support the rider.

Example: 1 The handlebar steers the airboard.

Vocabulary   **8**   Match the pictures with the sentences.

1 X is attached to Y.   3 X is mounted on Y.
2 X is suspended from Y.   4 X is connected to Y.

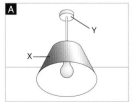   

9   Complete these sentences. Use each phrase once only.

attached to   connected to   mounted on   suspended from

1 The huge cables of the Millau Bridge are _____ steel pylons.
2 The pylons and the road deck are _____ concrete piers.
3 Close the circuit switch. Now the lamp is _____ the current.
4 The shelf is _____ the wall with screws.

2 Hotline

Listening 1 ▶🎧 48 Listen to the automated message on the phone. The customer wants to talk to the service technician about a computer problem. Which three keys does the customer press?

2 ▶🎧 49 The customer gets through to the service technician. What does the technician say? Complete the text below.

- Hello, you've (1) _____ the computer service hotline. This is Jan (2) _____. I'm the technician. How (3) _____ I (4) _____ you?

3 ▶🎧 50 Listen to this phone call to a service hotline. What mistakes did the customer make when he set up his wireless router? Delete the wrong words.

1 The router *is/isn't* connected to the *power outlet/computer/modem*.
2 The customer *has/hasn't* connected the computer to the *power outlet/router/modem*.

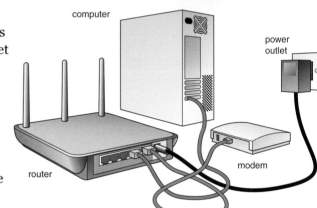

Speaking 4 Work in pairs. Practise similar conversations.

- *Hello, is that the IT hotline?*
- Yes, this is … speaking. I'm the technician. How can I help you?
- *My router doesn't work.*
- OK. I'll talk you through it. Are you sitting at the computer now?
- *Yes, I am.*
- OK. Look at the back. Is the … connected to the …?

USEFUL LANGUAGE

Is the … connected to the …?
Have you connected your … to the …?

5 Work in pairs. Make more dialogues about the situations in these pictures.

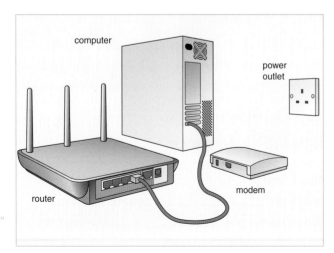
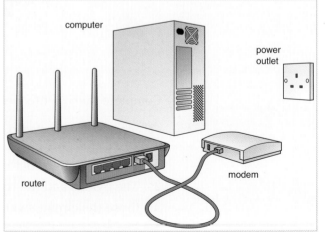

9 | Troubleshooting

Language

6 Write short form answers for these questions.

1. Are the lights on? ✓ *Yes, they are.* ✗ *No, they aren't.*
2. Is the computer connected to the adapter?
   ✓ _____ ✗ _____
3. Have you sent the email?
   ✓ _____ ✗ _____
4. Does your new radio work?
   ✓ _____ ✗ _____
5. Did you go to the cinema yesterday?
   ✓ _____ ✗ _____
6. Can I speak to your brother?
   ✓ _____ ✗ _____
7. Do you work in the city?
   ✓ _____ ✗ _____
8. Are you sitting at the computer now?
   ✓ _____ ✗ _____
9. Do those speakers cost a lot of money?
   ✓ _____ ✗ _____
10. Has your car broken down?
    ✓ _____ ✗ _____

7 ▶ 51 Look at 6 again and listen to the questions and answers. You will hear only one answer to each question. Repeat each answer.

Task

8 Work in pairs. Find out all the differences between your wiring diagram and your partner's.

Hint: there are at least ten differences of (a) location of sockets and (b) wiring connection.

Instructions.
- Student A, turn to page 117.
- Student B, this is your wiring diagram.

USEFUL LANGUAGE

digital receiver, DVD, VCR, TV, antenna, SCART socket, RF socket, in, out, power, socket

Do you have a/an ... ?
Look at the ... .
Where is the ... ?
Does the ... connect to the ... ?
Have you connected the ... to the ... ?
Is the ... connected to the ... ?

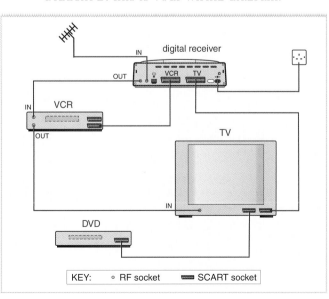

Troubleshooting 9

3 User guide

Start here 1 🔊 52 Listen and complete the flow chart.

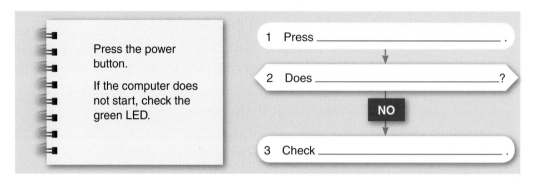

Press the power button.

If the computer does not start, check the green LED.

1 Press _____ .
2 Does _____ ?
   NO
3 Check _____ .

Reading 2 Draw a similar flow chart based on the solutions in this troubleshooting guide.

Notebook computer – troubleshooting FAQ

I pressed the power button and opened the display, but the computer does not start or boot-up.

Try these solutions:
1 Press the power button again.
2 If the computer does not start, check the green LED.
3 If the green LED is off, check the power source.
4 If the power source is off, switch on the power and press the power button again.
5 If the computer does not start, check the disk drive.
6 If there is a disk in the drive, take it out and press the power button again.

Language

Condition		Instruction
If	the car starts,	drive away.
	the car doesn't start,	check the battery.
	the light is off,	press the power button.
	there is a disk in the drive,	take it out.

3 Make sentences with *if* from these short dialogues.

1 ● Is the light on?
  ● No, it isn't.
  ● OK. Press the switch.

2 ● Does the airboard start?
  ● No, it doesn't.
  ● OK. Turn the key.

3 ● Are there any numbers on the screen?
  ● No, there aren't.
  ● OK. Press the keys.

4 ● Are the LEDs off?
  ● Yes, they are.
  ● OK. Push the power button.

5 ● Is the battery flat?
  ● Yes, it is.
  ● OK. Either replace it or recharge it.

6 ● Do the speakers work?
  ● Yes, they do.
  ● OK. Connect them to the computer.

Example: 1 If the light isn't on, press the switch.

4 Draw a flow chart. Use the information from the text.

Turn the key. If the car starts, drive away. But if the car doesn't start, check the battery. If the battery doesn't work, recharge it. If the battery works, check the starter motor.

Writing

5 Write a troubleshooting guide based on this dialogue. Write six sentences.

- *Hello, service hotline here, Mike speaking. How can I help you?*
- Hello. I've got a problem with my printer. It doesn't print.
- *OK. First check the cable between the printer and your computer. Is it loose?*
- Yes, it is.
- *OK. Connect the cable. Now check the power. Is the printer on?*
- Yes, it is.
- *Right. Now try to print. Is it printing?*
- No, it isn't.
- *OK. Now check the paper. Is there any paper in the printer?*
- No, there isn't.
- *OK. Put some paper in the printer. Now try to print again. Does it print?*
- No, it doesn't.
- *All right. Switch off and wait for ten seconds. Then switch on again.*
- It's printing! Thanks for your help.
- *You're welcome. Goodbye.*

Begin:

1 *If you can't print, check the cable between the printer and the computer.*

2 *If the cable is loose, connect … and check … .*

Social English

6 Complete the dialogues with short answers.

1 ● *Do you live near here?*
  ○ _____. I live less than a kilometre away.

2 ● *Do you work at BMW?*
  ○ _____. I work at Mercedes.

3 ● *Are you in IT?*
  ○ _____. I'm in engineering.

4 ● *Have we met before?*
  ○ _____. We met at the conference.

5 ● *Did you drive here?*
  ○ _____. I came by train.

7 Work in pairs. Practise the dialogue in 6.

8 Work in pairs. Make similar dialogues, using the information below.

more than 20 miles away / Citroën + Renault / R&D + quality control / in Paris / cycle + bus

10 Safety

1 Rules and warnings

Start here 1 Work in pairs. What safety rules are in your workplace or college? Make a list.

2 🎧 53 Listen and complete the warnings with the words in the box.

don't   might   must   mustn't

1 You _____ wear a hard hat on the building site.
2 _____ go through that door!
3 You _____ wear safety gloves everywhere in the factory.
4 _____ touch that machine! It's very hot.
5 Be careful! High-voltage electricity. You _____ get an electric shock.
6 You _____ use your mobile phone here.

Reading 3 Work in pairs. Why do the signs below have different colours and shapes?

4 Read the text. Match the examples to the signs.

The safety signs below follow the ISO international standard.
This standard is used in the EU because it has many different languages.
There are three types of safety sign:

- **WARNING SIGNS.** These signs warn you about a danger. They say things like this: *Warning. Danger. Be careful. Look out. There is a danger or hazard here. You might injure yourself.* The signs are yellow and black in colour and triangular in shape. Here are some examples:
  1 Warning. Poison: see (1) _C_
  2 Danger. Fire hazard here: see (2) _____

- **PROHIBITION SIGNS.** These signs prohibit an action. They say: *Do not do this. You must not do this. Never do this.* The signs are red, white and black in colour and round in shape. Here are some examples:
  3 You must not lift this with a hook: see (3) _____
  4 Never take the guard off this machine: see (4) _____

- **MANDATORY ACTION SIGNS.** These signs order you to do something. They say: *Do this. You must do this. Always do this.* These signs are blue and white in colour, and round in shape. Here are some examples:
  5 Always read the manual before you service this machine: see (5) _____
  6 You must use the guard on this circular saw: see (6) _____

Language

	Wear	a hard hat here.	Do not / Don't	touch the machine.
Always	wear		Never	
You must			You must not / You mustn't	

5 Complete the instructions with the words in the box.

| always | do | do not | must | mustn't | never |

1 _____ use a lighted match in this workshop.
2 _____ wash your hands after using these chemicals.
3 _____ enter this small space.
4 You _____ wear safety boots when you lift this.
5 _____ not smoke in this factory.
6 You _____ touch this machine with bare hands. It's hot.

6 Write these signs in another way.

Example: 1 Do not smoke here.

No smoking | No mobile phones | No running | No entry | No exit | NO PARKING

Use *might* or *could* to explain the possible result of the hazard.

| You | might / could | burn your arm. injure/hurt yourself. get an electric shock. |

7 Complete these warnings with the words or phrases in the box. You can use the words or phrases more than once.

| could | might | there are | there's |

1 Take care. Heavy weight. You _____ injure your back.
2 Warning. _____ a cold surface here. You _____ injure your hands or arms.
3 Be careful. You _____ trap your hand in the gears.
4 Danger. _____ lasers in this laboratory. You _____ injure your eyes.

2 Safety hazards

Start here   **1**   🔊 54   Listen and match the warnings with the pictures.

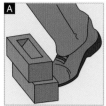  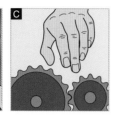  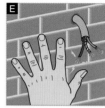 

2   Listen again and write the warning number in the table.

Warning	Possible result
	You might burn your hands.
	You could injure your head.
	You might fall into the gap.
	You could trip over the bricks.
	You might trap your hand in the gears.
	You could get an electric shock.

Speaking   **3**   Say the warnings and their possible results.

Example: 1 Look out! There's a low beam in front of you. You could injure your head.

4   Work in pairs. How many safety hazards can you see? Make a list.

10 | Safety

5 You are a safety inspector, inspecting the workshop in 4. Describe what you see.

There is	a	liquid	in the workshop.	A cable	is	damaged.
There's	some	bricks	on the floor.	Two windows	are	locked.
There are	no	boxes	around the bricks.	The fire exit		broken.
		food	on the machines.	Some cables		coiled.
		drink	on the stairs.			
		tools	on the benches.			
		fire extinguishers				
		fire exit				
		cones				
		guards				

Language Past simple of *is* and *are*.

	There was	some liquid	on the floor.
	There were	some boxes	on the stairs.
The fire exit	was		locked.
Some cables	were		coiled.

6 Change more sentences from 5 into the past.

Writing **7** Complete the inspector's report. Describe all the hazards in the workshop.

Safety inspection report

Visit to: Kwik Automotive Workshop
Date of report: 25th October

I inspected the workshop on 22nd October. Here are my findings.
1 There were no fire extinguishers anywhere in the workshop.
2 There was a single fire exit, but the door was locked with a padlock.
3

8 Work in small groups. Write at least ten safety rules for the workshop in 4.

Put away all tools after work.
Do not bring food or drink into the workshop.
No eating or drinking in the workshop.
Always … .
Never … .
Staff must/must not … .

3 Investigations

Start here 1 Work in pairs. Discuss these questions.

- What's happening?
- Which directions are the planes moving in?
- Who will talk to the pilots?

2 🔊 55 Listen and complete the warning to the pilot from air traffic control.

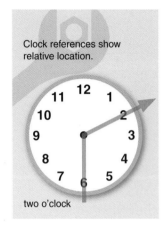

Clock references show relative location.

two o'clock

- ● ConAir 286. Unknown traffic. (1) _____ o'clock. (2) _____ metres. Crossing right to left.
- ○ ConAir 286. Negative contact. Request vectors.
- ● Turn (3) _____. Heading (4) _____. Descend. (5) _____ metres.
- ○ Right turn. Heading (6) _____. Descending. (7) _____ metres. ConAir 286. …
- ● Con Air 286. All clear. Resume own navigation.
- ○ Roger. ConAir 286.

Reading 3 Read this newspaper article and complete the incident report.

Near Miss Over Manchester

25 November

Last night, a military jet plane almost crashed into a large passenger plane over northern England.

The incident happened in dense clouds 10 km west of Manchester. The Boeing 757 passenger plane was 3505 metres above sea level. At 22.17, the F16 military plane passed at an altitude of 3527 metres. At its closest point, the total distance between the two aircraft was only 36 metres.

The Boeing, flight number BA 4058, had 234 passengers, and was on a flight path from Manchester to Greece. The military plane was on its way from Scotland to the south of England.

The pilot and passengers on the plane did not see the incident because of the clouds, but the emergency anti-collision system (TACS) in BA 4058 switched on automatically. The TACS system steered the passenger plane safely away from the military plane.

There were no injuries in the incident.

Aviation near-miss incident report

Date of incident:
Time:
Location:
Distance between two planes:

PLANE 1
Type: *Boeing 757 passenger plane*
Altitude:
Flight number:
Number of passengers:
Flying from:
Flying to:

PLANE 2
Type:
Altitude:
Flight number: –
Number of passengers: *none*
Flying from:
Flying to:

78 | 10 | Safety

Speaking   **4**   Work in pairs: an investigator and a pilot. Ask and answer these questions.

1 Where / incident / happen
2 When / it / take place
3 How high / be / Boeing
4 What / be / height / of / F16
5 What time / F16 / pass / Boeing
6 How far / be / jet / from / passenger plane
7 What / be / flight number / passenger plane
8 How many passengers / be / in / Boeing

take place = happen

Language

Where	were	the planes?		(They were) 3500 m above NW England.
When	did	the incident	happen?	(It happened) at 22.17.

Task   **5**   Work in pairs. Follow the instructions.

Student A. Turn to page 115.

Student B:
1 Investigate Student A's incident. Ask questions and complete the report form.
2 Change roles. Your incident is on page 118.

About the accident	About the injured person
Date: _____	Name: _____
Time: _____	Job title: _____
Location: _____	Injury: _____
Height above ground: _____	**Description of accident**
Type of accident (tick one box):	
• lifted something and injured self ☐	
• received an electric shock ☐	
• slipped, tripped or fell on the same level ☐	
• fell from a height ☐	
• other ☐ _____	

Social English   **6**   Complete the dialogue with the words in the box.

     are   can't   don't   I'd   I'll   must

● *We (1) _____ go out for a drink soon.*
○ Yes, (2) _____ like to do that. How about tomorrow? (3) _____ you free tomorrow?
● *I'm sorry, I (4) _____ do it tomorrow. What about Saturday?*
○ Yes, Saturday's fine. What time?
● *I (5) _____ know yet. (6) _____ phone you tomorrow morning.*
○ OK, good. Talk to you then.

7   Work in pairs. Practise the dialogue in 6 with your partner.

8   Work in pairs. Make similar dialogues. Use different times and days.

go and see a film / have a meal together / go bowling / have a party

Safety **10**

Review Unit E

1 Complete the sentences with the correct forms of verbs in the box.

control	increase	move
propel	push	rotate	steer
support	turn

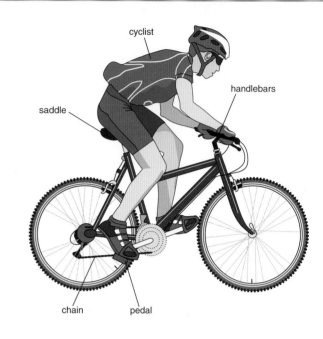

1 The saddle _____ the cyclist. The cyclist _____ the pedals downwards.
2 The pedals _____ the chain and the wheels _____. This _____ the bike forwards.
3 The cyclist uses the pedals to _____ the speed. If the cyclist pedals quickly, this _____ the speed of the bike.
4 The cyclist _____ the bike with the handlebars.
5 If the cyclist _____ the handlebars to the left, the bike goes left.

2 Complete the description with the correct form of the verbs in the box.

| contain | drive | move | suck | work |

This hovercraft moves over land and water. How does it (1) _____? A powerful engine (2) _____ two large fans. The fans (3) _____ the air in. They force some of the air backwards and push some of the air downwards. A rubber skirt (4) _____ the air and the hovercraft (5) _____ on the cushion of air.

3 Complete the sentences with the words and phrases in the box.

| above | below | between | in the centre | on the left/right | to the left/right |

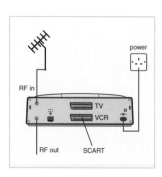

1 The RF sockets are _____.
2 The SCART sockets are _____.
3 The SCART sockets are _____ the RF sockets and the power socket.
4 The power socket is _____ of the SCART sockets.
5 The RF OUT socket is _____ the RF IN socket.
6 The TV SCART socket is _____ the VCR SCART socket.

80 Review Unit E

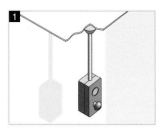

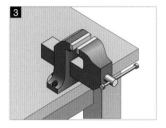

4 Identify the equipment from the description. Use the words in the box.

| battery | digital receiver | disk drive | modem | router | starter motor |

1 This device can change digital signals into analogue signals for a TV.
2 This device stores electricity. When it is flat, you recharge it.
3 This equipment can connect two or more computers to one modem.
4 This device connects a computer to the Internet through a phone line.
5 This machine uses electricity from a battery. It starts the engine of a car.
6 This hardware can copy data from a computer to a CD-ROM.

5 Look at the pictures and complete the sentences with the phrases in the box. You can use the words more than once.

| attached to | connected to | disconnected from | mounted on | suspended from |

1 The switch is _____ the ceiling.
2 The printer is _____ the power socket.
3 The vice is _____ the workbench.
4 The mouse is _____ the computer.
5 The hook is _____ the rope. The rope is _____ a bar.
6 The speaker is _____ a base. It is _____ the computer.

6 Draw and complete the flowchart.

If your computer does not start, check the adapter. If the adapter is not connected, connect it to the computer. If the adapter is connected, check the disk drive. If there isn't a disk in the the drive, press the power button. If there is a disk in the drive, take it out.

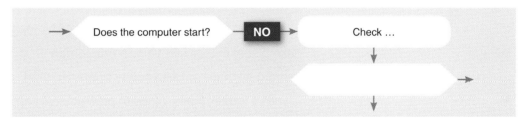

7 Write a troubleshooting guide from this flowchart.

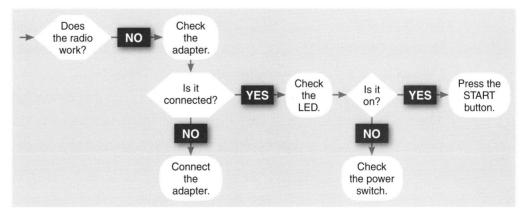

Begin:
If the radio doesn't work, check the adapter

8 Complete the warnings. Use each word once only.

| always | could | don't | might | must | mustn't | never |

1 Staff _____ wear hard hats at all times on this site.
2 You must _____ use a lighted match near petrol or gas.
3 You _____ smoke in the workshop or on the building site.
4 This low beam is very dangerous. You _____ injure your head on it.
5 _____ wear gloves if you lift these boxes. They have sharp edges.
6 The oven is very hot. You _____ burn yourself. Please _____ touch it.

9 Complete the safety report with the correct form of the verbs in brackets.

On 24th August last year, I inspected the Nautilus shipyard. I (1) _____ (find) many safety hazards. Here are the main points of my safety report.

The emergency exit (2) _____ (be) locked. There (3) _____ (be) some ropes on the ground, between two boats. Two fire extinguishers (4) _____ (be) damaged. Five workers (5) _____ (have) no hard hats or safety gloves. One welder (6) _____ (not wear) his safety boots. A high-voltage cable (7) _____ (be) coiled. There (8) _____ (be) many tools on the ground.

A supervisor (9) _____ (tell) me about a near miss. The incident (10) _____ (take place) in July last year. A repair man (11) _____ (put on) his hard hat and safety boots. He then (12) _____ (climb) a ladder 8 metres up to an electrical cable. The cable (13) _____ (be) damaged. It (14) _____ (have) some bare wires. The repair man (15) _____ (shout) to a worker: 'Switch off the power!' The worker (16) _____ (switch off) the main electricity supply and shouted: 'OK, I've (17) _____ (switch) it off!' Then the repair man (18) _____ (touch) the cable. But the cable (19) _____ (not be) connected to the mains supply. It (20) _____ (be) connected to a generator. There (21) _____ (be) a spark. The repair man was very lucky. He (22) _____ (not receive) a shock. But this was a very serious incident.

Review Unit E

10 Ask the questions for these answers about the near miss incident in 9.

1 It took place in the Nautilus shipyard. (Where / incident)
   Where did the incident take place?

2 It happened in July last year. (When / happen)

3 Yes, he wore his hard hat and his safety boots. (repair man / hard hat)

4 He used a ladder. (How / climb / to the cable)

5 It was about 8 metres high. (How / cable)

6 It had some bare wires. (problem)

7 No, he didn't, but there was a spark. (get / electric shock)

8 No, it wasn't. It was connected to a generator. (cable / mains supply)

11 Write a set of safety rules based on the report in 9.

Project **12** Choose one of the projects below and follow the instructions.

1 Troubleshooting in your industry
   Work with a partner or small group from the same or similar industries.
   a) Find out about some important equipment in your industry.
   b) Make a list of common operating problems, and their solutions.
   c) Write a troubleshooting guide explaining how to solve the problems.

2 Safety in your industry
   Work with a partner or small group from the same or similar industries.
   a) Find out about the causes of common accidents in your industry.
   b) Design a safety poster to avoid one of these accidents.
   c) Write a set of safety rules for your poster.

11 Cause and effect

1 Pistons and valves

Start here 1 Put the parts of the spray bottle together. Draw arrows to show where the parts fit the bottle.

Turn to page 113 to check your answers.

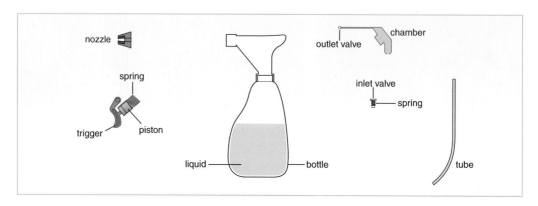

2 Work in pairs. How does the pump in the spray bottle work? Discuss with your partner.

Reading 3 Match each diagram with a caption below.

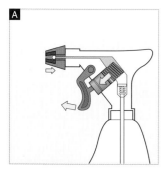
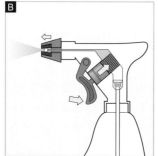
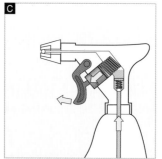
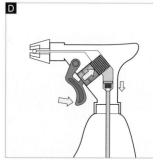

increase the temperature/pressure/speed/volume

decrease the temperature/pressure/speed/volume

Caption 1: The trigger makes the piston move in. This makes the water pressure increase. The high pressure causes the outlet valve to open. The open outlet valve allows the water to flow out of the chamber.

Caption 2: The piston moves in. This causes the water pressure to increase. The high pressure makes the inlet valve close. The closed inlet valve prevents the water from flowing back into the bottle.

Caption 3: The piston moves out. This makes the water pressure decrease. The low pressure causes the inlet valve to open. The open inlet valve lets water flow from the bottle into the chamber.

Caption 4: The piston moves out. This makes the water pressure decrease. The low pressure causes the outlet valve to close. The closed outlet valve stops air from flowing into the chamber.

Language

The motor	causes	the shaft	to	move.
	makes	the shaft		move.
The open valve	lets	the water		flow out.
	allows	the water	to	flow out.
The closed valve	prevents	the water	from	flowing out.
	stops			

4 Make true sentences about the pump.

| The trigger
The piston
The spring
The two valves
The inlet valve
The outlet valve
High pressure
Low pressure | make(s)
let(s)
cause(s)
allow(s)
prevent(s)
stop(s) | the water
the piston
the inlet valve
the outlet valve
the piston
the pressure
the air | (to)
(from)
(–) | flow in/out/back.
flowing in/out/back.
move in/out/in and out.
increase.
decrease.
open.
close. |

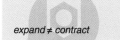

expand ≠ contract

5 Rewrite these sentences to give similar meanings. Replace the verb(s) in *italics* with the correct form of the verb(s) in brackets.

1 Heat makes a metal expand and cold *makes* it contract. (cause)
2 Overflow pipes *let* extra water flow out of the tanks. (allow)
3 The valve on the oil well *does not allow* the oil to explode. (prevent)
4 These powerful pumps *force* the water to rise 30 m up the hill. (make)
5 These fire extinguishers *do not allow* electrical fires to spread. (stop)
6 Show your ID card and the guard will *allow* you to enter the factory. (let)

6 Delete the wrong words.

PISTON PUMPS

Piston pumps can pump any fluid. This one pumps water. The pump has a motor, a shaft, a piston, a spring and two valves. The valve on the right is the outlet valve. The valve on the left is the inlet valve.

This is how it works. The motor makes the shaft (1 move/to move) in and out. The shaft makes the piston (2 move/to move) in and out. Let us look at the two movements of the piston.

1 The piston moves in. This causes the water pressure (3 increase/to increase). The high pressure forces the outlet valve (4 open/to open). The open valve allows the fluid (5 flow/to flow) out of the pump through the outlet pipe. At the same time, the high pressure makes the inlet valve (6 close/to close). This closed valve prevents the fluid (7 to flow/from flowing) back through the inlet pipe.

2 The piston moves out. This makes the water pressure (8 decrease/to decrease). The low pressure forces the inlet valve (9 open/to open). The open inlet valve lets fluid (10 flow/to flow) into the pump through the inlet valve. At the same time, the low pressure makes the outlet valve (11 close/to close). This closed valve stops the fluid (12 to flow/from flowing) back into the pump through the outlet pipe.

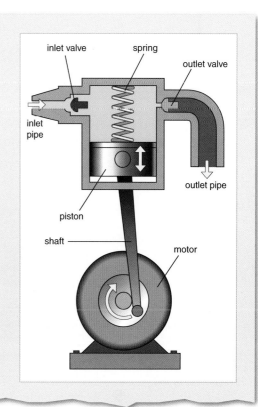

2 Switches and relays

Start here 1 Work in pairs. Try this quiz. How many electrical symbols do you know?

battery, bell, buzzer, conductor, earth, lamp, switch, terminal

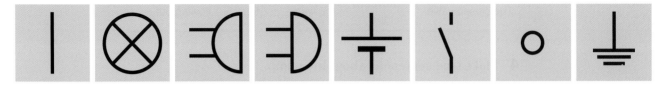

Answers: see the glossary of electrical symbols on page 109.

2 🔊 56 Listen and name the sounds. Choose from the list below.

buzzer, door bell, click, siren, horn, beep, alarm bell, dial tone

Reading 3 Work in pairs. How does this window burglar alarm work?

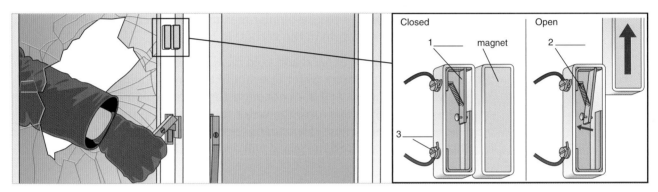

4 Read the web page. Label the circuit diagram and the diagram in 3.

battery   buzzer   spring   switch   terminal   wire

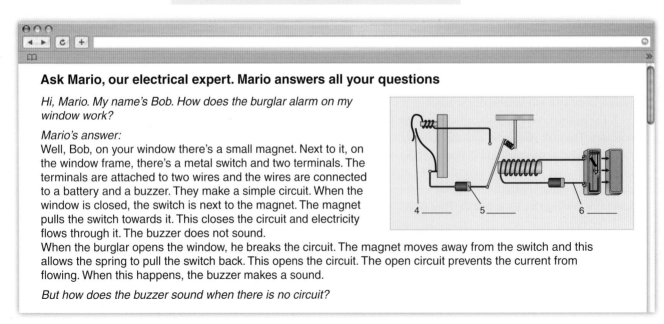

Ask Mario, our electrical expert. Mario answers all your questions

Hi, Mario. My name's Bob. How does the burglar alarm on my window work?

Mario's answer:
Well, Bob, on your window there's a small magnet. Next to it, on the window frame, there's a metal switch and two terminals. The terminals are attached to two wires and the wires are connected to a battery and a buzzer. They make a simple circuit. When the window is closed, the switch is next to the magnet. The magnet pulls the switch towards it. This closes the circuit and electricity flows through it. The buzzer does not sound.
When the burglar opens the window, he breaks the circuit. The magnet moves away from the switch and this allows the spring to pull the switch back. This opens the circuit. The open circuit prevents the current from flowing. When this happens, the buzzer makes a sound.

But how does the buzzer sound when there is no circuit?

5 Work in pairs. What is the answer to Bob's second question?

6 Read the next part of the web page. Check your answer to 5.

Because there is another circuit. The buzzer has its own circuit. When the window circuit opens, this makes the buzzer circuit close.

How does this happen?

The buzzer circuit has its own battery, an electro-magnet and a relay switch. This is how it works:
1 The window circuit opens.
2 This causes the electro-magnet in the window circuit to switch off.
3 The electro-magnet releases the relay switch on the buzzer circuit. This allows the spring to push the switch. The buzzer circuit closes.
4 The current flows from the battery around the buzzer circuit. This makes the buzzer produce a loud noise.

OK, I understand the circuit. But how does the buzzer make a sound?

That's easy. Here's what happens:
1 The current flows through the buzzer circuit.
2 The current makes the electro-magnet switch on.
3 The electro-magnet pulls the metal strip away from the thin wire.
4 This causes the current to switch off again.
5 When the current switches off, the electro-magnet switches off.
6 This allows the metal strip to spring back towards the thin wire.
7 The metal strip moves quickly up and down. This makes the loud buzzing noise.

Thanks, Mario. I get it now.

---

7 Answer these questions about the complete burglar alarm.
1 How many circuits are there?
2 How many electro-magnets are there? What is an electro-magnet?
3 How many switches are there?
4 What makes each switch open and close?

Language

8 Complete the sentences with the correct form of the verbs in the box.

| allow | cause | let | make | prevent | stop |

1 The electro-magnet _____ the relay switch move away from the contact.
2 The magnet _____ the window switch from opening.
3 The wires _____ the electric current to flow from the battery to the electro-magnet.
4 The open switch _____ the current from flowing around the window circuit.
5 The spring _____ the window switch to break the window circuit.
6 The closed switch _____ the current flow around the buzzer circuit.

Speaking

9 Work in pairs. Explain how the burglar alarm works. Look at the circuit diagram, but don't look again at the reading text.

3 Rotors and turbines

Start here

1 Try this quiz. What do you know about wind turbines?

> 1. **How tall is the tower of the world's tallest wind turbine?**
>   a) about 100 m   b) about 180 m   c) about 200 m
> 2. **How high is the world's highest turbine?**
>   a) about 1800 m   b) about 2300 m   c) about 2600 m
> 3. **What's the minimum wind speed for a large wind turbine?**
>   a) about 15 km/h   b) about 20 km/h   c) about 25 km/h
> 4. **What's the maximum wind speed for a large wind turbine?**
>   a) about 45 km/h   b) about 70 km/h   c) about 90 km/h

2 🔊 57 Listen to this radio programme and check your answers to the quiz.

Vocabulary

3 Label this diagram with the parts of a wind turbine in the box.

> blade   brake   gear   generator   housing   hub   shaft

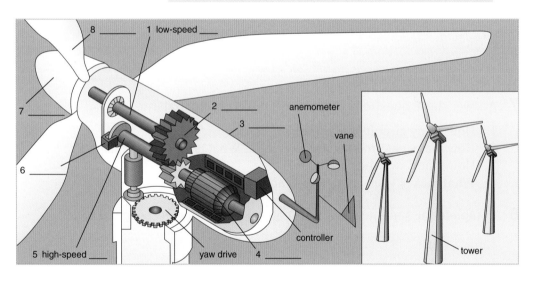

4 Read the text. Check your answers to 3.

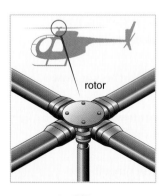

> The wind turbine consists of a tower, a rotor and a housing. The rotor consists of three blades, and a hub.
>
> The housing is a strong rigid container. It contains a low-speed shaft, a high-speed shaft, two gears, a generator, a controller, and a brake.
>
> The low-speed shaft connects the rotor to the gears. The high-speed shaft connects the gears to the generator.
>
> Inside the housing, at the back, behind the generator, is the controller.

11 | Cause and effect

Reading 5 Read the web page and answer the questions below.

TECHNO CHANNEL: the TV channel for people who love technology

Yesterday, Techno Channel interviewed the wind turbine expert, Dr Roger Jones. Here is part of the script. To download the whole script, click here.

How does the wind turbine work?

The wind blows on the blades and makes them rotate. This causes the shaft to rotate at a speed of about 30–60 rpm.

But isn't that too slow? The shaft in a generator must rotate at about 1200–1400 rpm.

That's right. There are two shafts. There's a low-speed shaft and a high-speed shaft. The low-speed one is attached to a large gear. The high-speed one is attached to a small gear. The large gear makes the small gear turn and the small gear makes the high-speed shaft rotate. This shaft rotates at 1200–1400 rpm.

Ah, I see. And it drives the generator at this speed?

That's right. And then the generator produces AC electricity.

What happens if the wind is too strong?

The anemometer measures the speed of the wind. It sends this data to the controller. (The controller is a small computer.) If the speed of the wind is more than about 90 km/h, the controller automatically switches off the wind turbine. This prevents the wind from damaging the turbine.

1 Which part makes the low-speed shaft turn?
2 What are the two main functions of the controller?
3 Which part transmits rotation to the generator?

data = information

6 What do these words refer to? Choose the correct answer.

1 *one* (line 10)   a) generator   b) shaft   c) gear
2 *it* (line 14)   a) low-speed shaft   b) high-speed shaft   c) small gear
3 *It* (line 17)   a) anemometer   b) speed   c) wind

Language 7 Complete the sentences with the correct form of the verbs in the box.

cause   make   prevent

1 The wind _____ the blades rotate.
2 The controller _____ the wind turbine from operating in a strong wind.
3 The blades _____ the low-speed shaft to rotate.

Speaking 8 Work in pairs. Explain how the wind turbine works. Look at the diagram, but don't look again at the reading text.

Social English You can use *let's* (= *let us*) to suggest something for you and others to do together.

Let's go to the café after work. Let's have a party for our class next week.

You can also say: *Why don't we go to the café after work? Why don't we have a party next week?*

9 Make your own suggestions.

1 A: *We have a free period after this class.*
   B: Let's _____.
2 A: *Work finishes early today.*
   B: Why don't we _____?
3 A: *Next week is the half-term holiday.*
   B: _____.
4 A: *The cinema is closed, so we can't see the film.*
   B: _____?

Cause and effect | 11

12 Checking and confirming

1 Data

Start here 1 Work in pairs. You are a TV reporter. Prepare questions about the Mars rover.

Reading 2 Read the text quickly. Does the text answer any of your questions?

include ≠ exclude
Weight of boat = 1000 kg.
This *excludes* crew, passengers and fuel.
Weight of crew, passengers and fuel = 200 kg.
Total weight of boat = 1200 kg.
This *includes* crew, passengers and fuel.

range = from minimum to maximum

Use *mass* on Mars, not *weight*.
If you travel to Mars, your weight changes, but your mass stays the same.

The Mars Science Laboratory, or MSL, is a rover, or mobile robot. It can move around on the surface of Mars.

Look at the diagram of the rover. It has a body, six wheels, two robot arms, two antennas and a mast. The antennas and the mast are mounted on the body, and the robot arms are attached to the front of the body.

There are special tools at the end of each robot arm. Some tools break pieces of rock. Other tools dig and collect samples of soil. Scientific instruments in the rover then analyse the soil and rock powder.

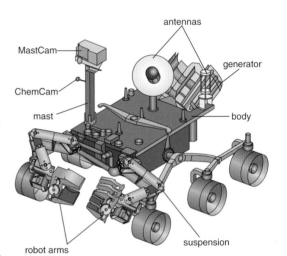

The top of the mast is about 2.1 metres above the ground. The mast supports two special cameras. They are called the MastCam and the ChemCam. The MastCam (mast camera) is at the top of the mast. It looks all around the rover. The ChemCam (chemistry camera) has a laser gun. The gun fires a laser beam at rocks up to 10 metres away and breaks them into powder. The camera then analyses the powder.

The rover is about 2.2 m long and its total mass is just under 800 kg. This includes at least 60 kg of scientific instruments.

It has a six-wheel drive and a special suspension system. The wheels are made of titanium and are 25 cm in diameter. The suspension system allows the six wheels to remain on the ground all the time. It also allows the rover to go over big rocks (up to 75 cm high), and over deep holes. Each wheel has its own motor. This allows the vehicle to rotate 360 degrees. It can move at a speed of up to 90 metres per hour. The average speed is about 30 metres per hour.

The rover can operate in the temperature range on Mars. This ranges from −120°C minimum up to 85°C maximum.

The rover can travel up to 200 metres per day and can operate for up to one Mars year (approximately 687 days).

3 Read the text again and complete this specification chart.

Mars Science Laboratory (Mars rover): specifications			
1 Total height		7 Maximum rotation of rover	
2 Total length		8 Maximum obstacle height	
3 Total mass		9 Maximum speed	
4 Mass of instruments		10 Average speed	
5 Number of wheels		11 Max./Min. temperature range	
6 Wheel size		12 Maximum daily distance	

Vocabulary Ways to express approximation:

| ~ about, approximately | > more than, over | ≤ up to |
| | < less than, under | ≥ at least |

4 Complete the sentences. Use the information in brackets.

1 The Mars rover _____.
 (height ~ 2.1 m; length ~ 2.2 m)
2 The rover _____.
 (mass > 750 kg)
3 The scientific instruments _____.
 (mass ≥ 60 kg)
4 The wheels _____.
 (rotation ≤ 360°)
5 The rover _____.
 (distance > 100 metres per day; operation ≤ ~ 687 days)

Speaking

5 Write questions for these answers about the rover.

1 It's called the Mars Science Laboratory.
2 It has six wheels.
3 Titanium.
4 They're attached to the front of the body.
5 It's mounted on the top of the body.
6 About 2.1 metres.
7 It looks at the whole area around the rover.
8 It fires a laser beam at rocks and analyses them.
9 Around 60 kilograms.
10 Up to 90 metres per hour.

6 Work in pairs. Practise asking and answering the questions in 5.

7 Work in pairs. Student A guess the answers. Then check them with Student B.
1 The diameter of Mars is … a) ~ 4280 km. b) ~ 6740 km. c) ~ 11,290 km.
2 Mars rotates 360° in … a) ~ 24 hours. b) ~ 36 hours. c) ~ 48 hours.
3 Mars is … kilometres from the Sun. a) ~ 220 million. b) ~ 150 million. c) ~ 300 million.
4 Mars orbits the Sun in … a) ~ 365 Earth days. b) ~ 685 Earth days. c) ~ 905 Earth days.

Example: 1 The diameter of Mars is about 4280 km. Is that right?

Student B: Turn to page 113.

Checking and confirming | 12

2 Instructions

Start here   **1**   Make a list of the instructions to give the Mars rover.

2   🎧 58   Listen and complete the dialogue between the controller and the rover.

- Move forwards 200 cm.
- ○ *Confirmed. I'm (1) _____ forwards 200 cm.*
- Now rotate 15 degrees to the left.
- ○ *Confirmed. I'm (2) _____ 15 degrees to the left.*

3   You are the rover. Confirm your actions.

Instruction	Confirmation
1 Move forwards 200 cm.	*I'm moving forwards 200 cm.*
2 Rotate 15 degrees to the left.	
3 Reverse for 300 cm.	
4 Rotate 80 degrees to the right.	
5 Go up the hill.	
6 Roll down the hill.	
7 Go round to the left of the rocks.	
8 Stop.	

Listening   **4**   🎧 59   Listen and complete the dialogue.

A is training B how to control the Mars rover.

A: *Right. I'll give you an instruction. First, do it. Then confirm what you're doing, OK?*

B: OK.

A: *Then confirm what the rover's doing. Is that clear?*

B: Yes.

A: *Right. Let's go. First, (1) _____ the rover (2) _____ 200 cm.*

B: OK. I'm (3) _____ the joystick forwards.

A: *Good. Now what's (4) _____?*

B: The rover (5) _____ moving.

A: *Right. Wait five seconds. Now what's happening?*

B: OK. It's (6) _____ forwards now.

Task   **5**   Work in pairs. Discuss the question below.

In this simulation on Earth, the Mars rover responds after five seconds. If the rover is on Mars, it responds after about ten minutes. Why?

Speaking   **6**   Complete the table. Use information from the table in 3 and the notes below.

Instruction	Confirmation	After 1 second	After 5 seconds
1 Make the rover move forwards 200 cm.	OK. I'm pushing the joystick forwards.	The rover isn't moving.	Now it's moving forwards.
2			
3			
4			

1   push joystick forwards
2   turn wheel left
3   pull joystick backwards
4   press 'rotate' button

7   Work in pairs. Practise the dialogues, using the notes in 3. Try not to look at the table.

Begin:
A: *Make the rover move forwards 200 cm.*
B: *OK. I'm pushing the joystick forwards.*
A: *Good. What's happening now?*
B: *The rover isn't moving.*
A: *That's OK. Wait for five seconds. Is it moving forwards now?*
B: *Yes, it is.*

8   Test your memory. Look at the pictures for 10 seconds. Then turn to page 113.

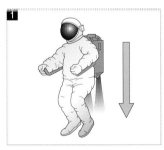

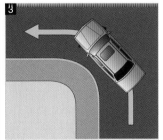

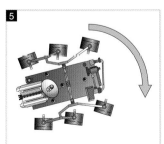
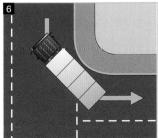
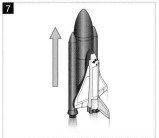
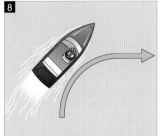

3 Progress

Start here 1 🔊 60 Listen to the astronaut talking about his work. Complete the list of tasks with the verbs in the box.

> assemble attach bring connect
> disconnect dismantle inspect
> remove repair replace take test

(1) _Test_ the equipment for the spacewalks.

On spacewalk 1: (2) _____ the damage.
(3) _____ photographs of the tank. Plan the repair and prepare for the next spacewalk.

On spacewalk 2: (4) _____ the pipes. (5) _____ the tank. (6) _____ the tank into the station.
(7) _____ the tank. (8) _____ the damage or
(9) _____ the part. (10) _____ the tank.

On spacewalk 3: (11) _____ the tank to the space station.
(12) _____ the pipes to the tank.

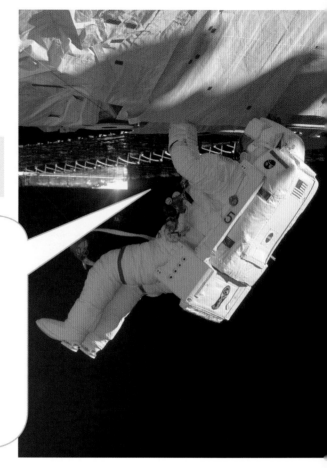

Vocabulary 2 Find the opposites of these words in 1.

connect, assemble, damage, remove

Listening 3 🔊 61 Listen to the controller talking to the astronaut. Complete the dialogue.

The controller is speaking from the control centre on Earth. The astronaut is on a space station.

Task	June		
	5	6	7
Do first spacewalk.	▓		
Repair the oxygen tank.		▓	▓

- OK, today is the 6ᵗʰ of June, 7 pm in the evening. I'm checking progress on the space station. Have you (1) _____ the first spacewalk yet?
- Yes, we have.
- Good. When (2) _____ you do it?
- We (3) _____ the spacewalk yesterday, on the 5ᵗʰ of June.
- Right. And have you (4) _____ the oxygen tank yet?
- No, we haven't (5) _____ it yet. We're still (6) _____ it.
- When (7) _____ you finish it?
- We'll complete the job tomorrow morning.

94 12 Checking and confirming

Language You can use *yet* with some questions and negatives in the present perfect. It means *up to now*.

1 We *haven't repaired* the oxygen tank *yet*.
2 A: *Have you repaired* the oxygen tank yet? B: No, *not yet*.

Speaking 4 Work in pairs. Make similar dialogues. Today is 17th June.

Task	June											
	10	11	12	13	14	15	16	17	18	19	20	21
Test equipment for first spacewalk.	■											
Do first spacewalk.		■										
Take photograph of damaged tank.			■									
Inspect damage to tank.				■	■							
Remove tank.						■						
Repair tank.							■	■				
Replace tank.										■		
Dismantle old ventilation system.						■	■	■				
Lubricate moving parts on all fans.									■	■	■	
Install new valves on pumps.												■

Task 5 Work in pairs. Follow the instructions.

- Student A: Turn to page 117.
- Student B:
It's 8th August. You're doing a progress check. Ask Student A questions and complete your checklist.

Task	Y/N?	Notes
Dismantle old water system	Y	Completed 4th Aug.
Assemble new water system	☐	
Install water system	☐	
Test equipment for third spacewalk	☐	
Take video of damaged nose cap	☐	
Inspect damage to waste tank	☐	
Assemble new robot arm	☐	
Attach new robot arm	☐	

B: *Have you dismantled the old water system yet?*
A: *Yes, we have.*
B: *When did you complete the job?*

Checking and confirming 12

Review Unit F

1 Complete the sentences with the correct form of the verbs in the box.

| allow   cause   let   make   prevent   stop |

1. The water flows down onto the water wheel. This _____ the wheel turn.
2. The valve opens. This _____ the water flow in.
3. The valve closes. This _____ the water from flowing out.
4. The switch touches the contact. This _____ the electric current to flow.
5. The switch moves away from the contact. This _____ the electric current from flowing.
6. The water level rises. This _____ the float to rise.

2 Complete the driving instructor's words with the correct form of the verbs in brackets.

1. If you _____ (press) the accelerator pedal, this _____ (make) the car _____ (go) faster.
2. If you _____ (push) the brake pedal down, this _____ (cause) the car to _____ (stop).
3. If you _____ (pull) the parking brake up, this _____ (prevent) the car from _____ (move).
4. If you _____ (release) the parking brake, this _____ (allow) the car to _____ (move) again.

3 Complete the sentences with the correct form of the verbs in the box.

| close   flow   from   go down   open   rise   to |

1. You push the handle down. This makes the piston _____.
2. The piston rises. This makes valve B _____ and causes valve A _____.
3. Valve B closes. This prevents water _____ into the chamber.
4. Valve A opens. This allows water _____ into the chamber.
5. You pull the handle up. This causes the piston _____.
6. The piston goes down. This makes valve B _____ and causes valve A _____.

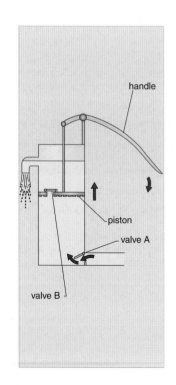

4 Draw a line from each word or phrase to its opposite.

increase   expand   bring   decrease   low   assemble
contract   dismantle   inlet   outlet   less than
take   more than   high   connect
remove   replace   disconnect   approximately   exactly

5 Complete this explanation of how the electric bell works with the correct form of the words in the box.

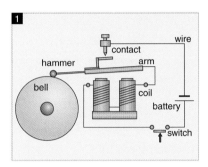 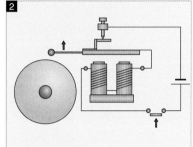

close   flow   make   move   open   pull   strike

How an electric bell works

Someone presses the bell button, and the switch (1) _____.
An electrical current (2) _____ through the coil. This
(3) _____ the coil become an electromagnet. The electromagnet
(4) _____ the metal arm towards it. (Diagram 1). This causes the
hammer to (5) _____ the bell. At the same time, it
(6) _____ the circuit. Now the coil is not a magnet. The hammer
(7) _____ away from the coil. (Diagram 2). This
(8) _____ the circuit again. The hammer (9) _____ the
bell again and again.

6 Work in pairs. Explain how this hand pump works.

7 Write your explanation of how the hand pump works.

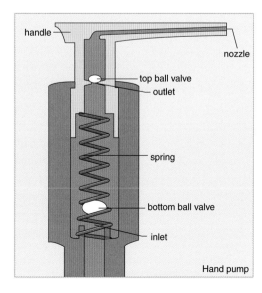

Hand pump

8 Complete these dialogues. Use the correct form of the verbs in brackets.

A supervisor in a car repair workshop is reporting on progress to his manager.

1 ● The men have _____ (replace) the windscreen.
  ○ Good. When did they _____ (replace) it?
  ● Let me check the file … They _____ (replace) it yesterday.

2 ● They've _____ (take) out the old brake system.
  ○ That's good. When did they _____ (take) it out?
  ● Let me see … They _____ (take) it out this morning.

3 ● Bob has _____ (drive) the car to the body repair shop.
  ○ That's great. When did he _____ (drive) it there?
  ● Let me check … Ah yes, he _____ (drive) it there about two hours ago.

4 ● Tom has _____ (speak) to the customer about the damage to her car.
  ○ Great. When did he _____ (speak) to her?
  ● Er, let me see … He _____ (speak) to her yesterday.

9 Work in pairs. Practise the dialogues in 8.

10 Work in pairs. Practise the dialogue below. The supervisor is checking progress with a mechanic. Then make new dialogues using the information from the table.

● Have you repaired the brakes yet?
○ Yes, I have.
● Good. When did you do that?
○ I did it yesterday.
● Right. And have you replaced the windscreen yet?
○ No, I haven't. I'm replacing it now.
● OK. And what about the main shaft? Have you lubricated it?
○ No, I haven't. I'll do that tomorrow morning.

Repair brakes	✓ yesterday
Replace windscreen	✗ in progress
Lubricate main shaft	✗ tomorrow morning

Lubricate axles and shafts	✓
Inspect damaged fuel tank	✓ last week
Disconnect fuel pipe from fuel tank	✓ yesterday
Take photographs of dented panels	✗ tomorrow morning
Remove old radiator	✗ tomorrow afternoon
Install new cooling system	✗ in progress
Repair dented bumpers	✓
Replace damaged valve on water pump	✗ in progress
Service the brake system	✗
Repair damaged radio	✗ later today
Connect battery to starter motor	✓ two days ago
Test new air conditioner	✓ 8.00 this morning

11 Write a description of this dam and how it works, using all the information and the words in the box.

| allow | carry | cause | drive | enter | flow | generate |
| leave | make | open | pass | produce | rotate | turn |

Function of dam
Main parts
Dimensions
Material
How the dam works
- gate / open → water / in
- water from reservoir → filter → gate → tunnel
- water → blades / turbine
- blades → shaft → generator
- generator → electricity
- high-voltage cables
- water / out

Hydroelectric dam

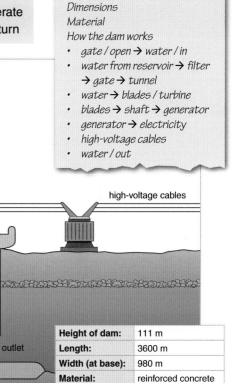

Height of dam:	111 m
Length:	3600 m
Width (at base):	980 m
Material:	reinforced concrete
Volume of water:	11,000 m³/s can pass through dam
Size of reservoir:	132 km³

Project 12 Find out some facts about a major engineering project in your country or region.

1. Draw a simple labelled diagram.
2. Make a specifications chart.
3. Write a short description of the project:
   - Function of project
   - Main parts
   - Dimensions
   - Materials
   - How it works

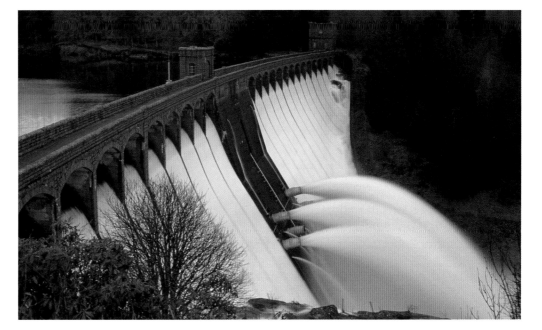

Review Unit F

Grammar summary

1 Present simple of *be*

Positive		
I	am	a student.
You	are	early.
He/She	is	a technician.
The machine (It)	is	on.
The switches (They)	are	off.
We/They	are	electricians.

I am ➔ *I'm*
you are, we are, they are ➔ *you're, we're, they're*
he is, she is, it is ➔ *he's, she's, it's*

Negative			
I	am	not	a technician.
You	are	not	late.
He/She	is	not	a student.
That	is	not	an M6 spanner.
We/They	are	not	from Italy.

I am not ➔ *I'm not*
you are not ➔ *you're not* or *you aren't*
he is not/she is not ➔ *he isn't/she isn't* or *he's not/she's not*
it is not ➔ *it isn't* or *it's not*
we are/they are ➔ *we aren't/they aren't* or *we're not/they're not*

Yes/No question		
Am	I	early?
Are	we	late?
	the switches	on?
	you	the manager?
Is	he/she	a technician?
	that	an AC adapter?

Don't use contractions in a short answer.
Are you French? Yes, I am. (Not ~~Yes, I'm.~~)
Is he a technician? Yes, he is. (Not ~~Yes, he's.~~)

Wh- question		
Where	are	we now?
Who	is	the manager?
	are	those men?
What	is	that sound?

In these tables, *Wh-* means any question word, e.g. *Where? When? How? How many? Why?*

What	is	that	called?
	are	those	called in English?

What is ➔ *What's*
You can say *What's this?* but not ~~*What's it?*~~
You have to say *What is it?*

2 Present simple of *have*

Positive		
I/You/We/They	have	25 screws.
My bike (It)	has	21 gears.

Negative			
I/You/We/They	do	not	have	any screws.
My bike (It)	does			27 gears.

does not ➔ *doesn't*
do not ➔ *don't*

Yes/No question			
Do	you/we/they	have	any screws?
Does	your bike (it)		27 gears?

In colloquial English:
Have you got any screws? (BrE) = *Do you have any screws?* (AmE)
I've got 25 screws. (BrE) = *I have 25 screws.* (AmE)

Wh- question				
How many	gears	does	your bike (it)	have?
	screws	do	you/we/they	

3 Present simple of other verbs

Positive

He/She	works	in Paris.
I/You/We/They	work	
This tool (It)	cuts	wood.
Those tools (They)	cut	

Negative

He/She	does	not	work	in Rome.
I/You/We/They	do			
This tool (It)	does	not	cut	metal.
These tools (They)	do			

does not ➔ *doesn't*
do not ➔ *don't*

Yes/No question

Do	you/they	work	in Paris?
Does	he/she		
Does	this tool (it)	cut	metal?
Do	these tools (they)		

Wh- question

Where	do	you/they	work?	
	does	he/she		
What	does	this tool (it)	do?	
	do	these tools (they)		

Spelling
There are three different ways to spell the ending of a present simple verb:

+ -s		+ -es		-y ➔ -ies	
flow	flows	go	goes	carry	carries
move	moves	pass	passes	study	studies
rise	rises	push	pushes	fly	flies

Pronunciation
There are three different ways to say the -s/-es ending of a present simple verb:

z	s	iz (rhymes with *his*)
flows	sinks	rises
moves	stops	passes
burns	strikes	presses
goes	hits	pushes

4 Modal verb: *can*

Positive

I/You/He/She/We/They	can	operate this machine.
A helicopter (It)	can	fly backwards.

Negative

I/You/He/She/We/They	can	not	operate the forklift truck.
An aeroplane (It)	can	not	fly backwards.

can not ➔ *can't* or *cannot*

Yes/No question

Can	I/you/he/she/we/they	operate this machine?
	a helicopter (it)	fly backwards?

Wh- question

How	can	I/he/she/we/they	help	you?
What	can	I/he/she/we/they	do	for you?

5 Modal verb: *will*

Positive and negative

I/You/He/She/We/They	will		build	the wall tomorrow.
	will	not		

will not ➔ *won't*
I will, you will, he will, she will, it will, they will ➔
I'll, you'll, he'll, she'll, it'll, they'll

Yes/No question

Will	I/you/he/she/we/they	build	the wall tomorrow?

Wh- question

When	will	I/you/he/she/we/they	build	the wall?

6 Modal verbs: *must*, *could* and *might*

You	must	wear a hard hat here.

You	must not / mustn't	touch the machine.

You	might / could	burn your arm. / hurt yourself.

7 Present continuous

Positive			
I	am	pressing	the brake pedal now.
You/We/They	are	breaking	the safety rules.
He/She	is	turning	the steering wheel.
The car (It)	is	moving	to the left.

I am → I'm
He is, She is, It is → He's, She's, It's
We are, They are → We're, They're

Negative				
I	am	not	pressing	the accelerator.
You/We/They	are	not	following	the safety rules.
He/She	is			
The car (It)	is	not	moving.	

is not → isn't
are not → aren't

Yes/No question			
Am	I	talking	to the manager?
Are	you/we/they	working	on the same project?
Is	he/she	wearing	a hard hat?
Is	your radio (it)	working?	

Wh- question			
Who	am	I	meeting today?
Why	are	you/we/they	leaving now?
Where	is	he/she	going?
What	is	the piston (it)	doing?

Spelling
There are three different ways to spell the *-ing* ending of a present continuous verb:

Add *-ing*		Lose final *-e* and add *-ing*		Double final letter and add *-ing*	
do	doing	leave	leaving	cut	cutting
go	going	move	moving	drop	dropping
break	breaking	rise	rising	put	putting

8 Present perfect

Positive			
I/You/We/They	have	damaged	the car.
He/She	has	broken	the windscreen.

I have, you have, we have, they have → I've, you've, we've, they've
he has, she has, it has → he's, she's, it's

Negative				
I/You/We/They	have	not	dented	the bumper.
He/She	has	not	broken	the lamps.

have not → haven't
has not → hasn't

Yes/No question			
Have	you/we/they	damaged	the car?
Has	he/she	broken	the windscreen?

Wh- question				
Where	have	you/we/they	parked	the car?
	has	he/she	driven	

9 Past simple

Positive			
I/We/They/He/She	went	to Madrid	last year.
The incident (It)	happened	last week.	

Negative				
I/You/He/She/We/They	did	not	go	to Paris last year.
The incident (It)	did	not	happen	yesterday.

did not → didn't

Yes/No question				
Did	I/you/he/she/we/they		go	to Paris last year?
	the incident (it)		happen	yesterday?

Wh- question				
When	did	I/you/he/she/we/they	go	to Madrid?
		the incident (it)	happen?	

Time expressions

Some time expressions you can use with the past simple:
- *yesterday, this morning, the day before yesterday*
- *three minutes ago, two days ago, five weeks ago*
- *last week, last month, last year*
- *in 2005, on the 20th October, at 6.30 am*

10 Past simple and past participle forms

The past participle is part of the present perfect verb. Here are some examples of verbs in this book.

Most verbs are regular. Both the past simple and the past participle end in *-ed*.

Regular (ending in *-ed*)	
verb	past simple/past participle
attach	attached
close	closed
connect	connected
cool	cooled
crack	cracked
crash	crashed
damage	damaged
dent	dented
disconnect	disconnected
drop	dropped
fit	fitted
happen	happened
inspect	inspected
launch	launched
mount	mounted
press	pressed
remove	removed
repair	repaired
replace	replaced
scratch	scratched
suspend	suspended
travel	travelled

Some verbs are irregular. The past simple and the past participle don't end in *-ed*.

Irregular (not ending in *-ed*) past simple = past participle	
verb	past simple/past participle
bend	bent
bring	brought
build	built
burn	burnt
buy	bought
cut	cut
find	found
get	got
have	had
hold	held
leave	left
let	let
lose	lost
make	made
put	put
read	read
say	said
sell	sold
send	sent
sit	sat
tell	told

Irregular (not ending in *-ed*) past simple ≠ past participle		
verb	past simple	past participle
become	became	become
break	broke	broken
do	did	done
drive	drove	driven
fall	fell	fallen
fly	flew	flown
go	went	gone
rise	rose	risen
run	ran	run
speak	spoke	spoken
steal	stole	stolen
take	took	taken
tear	tore	torn
write	wrote	written

Grammar summary

Pronunciation

There are three different ways to say the *-ed* ending of a past simple verb:

d	t	id*
flowed	launched	mounted
moved	increased	added
changed	dropped	inspected
opened	gripped	rotated

* rhymes with *did*

Here are some past participles often used as adjectives:

<u>Damage</u>
cracked, damaged, dented, punctured, scratched, broken, stolen, torn, bent, burnt, cut
<u>Location</u>
connected (to), disconnected (from), suspended (from), mounted (on), attached (to)

Example: *The pipe is cracked. The switch is connected to the battery.*

11 Past simple of *be*

Positive		
I/He/She	was	in London last year.
You/We/They	were	in the workshop yesterday.

Negative			
I/He/She	was	not	in Dubai last year.
You/We/They	were		in the workshop last week.

was not → *wasn't*
were not → *weren't*

Yes/No question		
Was	I/he/she	in Dubai last year?
Were	you/we/they	in the workshop last week?

Wh- question			
When	was	I/he/she	in London?
	were	you/we/they	in the workshop?

12 Zero conditional

If	the sun	shine	-s	,	the current flows from the panel.
	the sun	does not/ doesn't	shine	,	the current flows from the battery.

If	the battery	is	full	,	the current doesn't flow into the battery.
	the lamps	are not/ aren't	on	,	the current flows into the battery.

13 Countable and uncountable nouns

screws are countable			**cement** is uncountable	
a one	screw		some	cement
some two	screw	-s		
a bag of two bags of			a bag of two bags of	

Countable nouns can be both singular and plural.
Examples: *screw, nail, hammer, bottle.*
Uncountable nouns are always singular. Examples: *concrete, cement, sand, oil, water.*

How much/How many

Do you need	some/ any	screws? cement?	How	many	screws	do you need?
				much	cement	

14 Verb constructions

***cause, allow* + *to* infinitive**
***make, let* + bare infinitive**
***stop, prevent* + *from* + gerund**

The motor The open valve	causes allows	the shaft the water	to to	move. flow out.
The motor The open valve	makes lets	the water	flow out.	
The closed valve	prevents stops	the water	from	flowing out.

15 Describing damaged or missing items

Passive

The screen	is	scratched.
The speakers	are	

have/don't have

The cable	has	no	plug.
	doesn't have	a	
The cables	have	no	plugs.
	don't have	any	

There is/There are

There is	a scratch	on the screen.
	no manual	in the box.
There are	some scratches	on the screen.
	no batteries	in the box.

there is → there's
there are → there're

Reference section

1 Abbreviations

SI units of measurement

Abbreviations are usually *singular* (e.g. *50 metres* is *50 m*, not *50 ms*)

Abbreviations are usually *lower-case* (e.g. *mm*, not *MM*) with very few exceptions. Note that:
- *litre* can be *L* or *l*
- *ampere* (*A*), *watt* (*W*) and *volt* (*V*) use upper-case (capital) letters

Length

mm	millimetre(s)
cm	centimetre(s)
m	metre(s)
km	kilometre(s)

Area

mm²	square millimetre(s)
m²	square metre(s)
km²	square kilometre(s)

Volume/Capacity

mm³	cubic millimetre(s)
m³	cubic metre(s)
km³	cubic kilometre(s)
ml	millilitre(s)
cl	centilitre(s)
L (or l)	litre(s)

Mass/Weight

mg	milligram(s)
g	gram(s)
kg	kilogram(s)
t	tonne(s)

Electricity

A	ampere(s) or amp(s)
Ah	ampere hour(s)
W	watt(s)
kW	kilowatt(s)
kWh	kilowatt hour(s)
V	volt(s)

Speed

m/s	metre(s) per second
km/s	kilometre(s) per second
km/h	kilometre(s) per hour
rpm	revolution(s) per minute

Other units in common use

gal	*gallon(s)*	1 gal (US) = 3.7854 L
		1 gal (UK) = 4.5461 L
pt	*pint(s)*	1 pt (US) = 0.4732 L
		1 pt (UK) = 0.5683 L
in	*inch(es)*	1 in = 25.4 mm
yd	*yard(s)*	1 yd = 0.9144 m
mi (or m)	*mile(s)*	1 mi = 1.61 km
mph	*mile(s) per hour*	100 mph = 161 km/h
lb	*pound(s)*	1 lb = 0.4536 kg
oz	*ounce(s)*	1 oz = 28.3495 g

Temperature

°C	degree(s) Celsius
°F	degree(s) Fahrenheit

To convert Celsius to Fahrenheit:
°F = °C × 9/5 + 32.
To convert Fahrenheit to Celsius:
°C = (°F − 32) × 5/9.

Some other abbreviations used in this book

am	in the morning
AC	alternating current
approx.	approximately
CD	compact disc
CD-ROM	compact disc, read-only-memory
DC	direct current
DVD	digital video disc
etc.	and so on/etcetera
FAQ	frequently asked questions
GB	gigabytes
ID	identity
ISO	International Organisation for Standardisation
IT	information technology
LED	light-emitting diode
LH	left-hand
MB	megabytes
n/a	not applicable; write this when there is no possible answer, or no need to answer a question on a form
no.	number
NS	near-side (of car), away from the steering wheel
N, S, E, W, NW	north, south, east, west, north west
OS	off-side (of car), next to the steering wheel
pm	in the afternoon (or evening)
qty	quantity
R&D	research and development
ref.	reference/with reference to
RF	radio frequency; the RF IN socket on a TV comes from the antenna
RH	right-hand
SCART	a connector between two audio-visual machines, e.g. a TV and a DVD player, also called a Euro-connector
SI	International System of Units; metric units
TV	television
VCR	video cassette recorder

2 Numbers, times and dates

Numbers up to 100

1	one	14	fourteen
2	two	15	fifteen
3	three	16	sixteen
4	four	17	seventeen
5	five	18	eighteen
6	six	19	nineteen
7	seven	20	twenty
8	eight	21	twenty-one
9	nine	22	twenty-two
10	ten	23	twenty-three
11	eleven	24	twenty-four
12	twelve	25	twenty-five
13	thirteen		
30	thirty	70	seventy
40	forty	80	eighty
50	fifty	90	ninety
60	sixty	100	a hundred/one hundred

Numbers over 100

100	a hundred/one hundred
1000	a thousand/one thousand
10,000	ten thousand
100,000	a hundred thousand/one hundred thousand
1,000,000	a million/one million
1,000,000,000	a billion/one billion

Ordinal numbers

1st	first	11th	eleventh	21st	twenty-first
2nd	second	12th	twelfth	22nd	twenty-second
3rd	third	13th	thirteenth	23rd	twenty-third
4th	fourth	14th	fourteenth	24th	twenty-fourth
5th	fifth	15th	fifteenth	25th	twenty-fifth
6th	sixth	16th	sixteenth	26th	twenty-sixth
7th	seventh	17th	seventeenth	27th	twenty-seventh
8th	eighth	18th	eighteenth	28th	twenty-eighth
9th	ninth	19th	nineteenth	29th	twenty-ninth
10th	tenth	20th	twentieth	30th	thirtieth
				31st	thirty-first

Decimal numbers

0.1	nought point one/zero point one
15.1	fifteen point one
15.15	fifteen point one five
15.015	fifteen point oh one five/fifteen point zero one five

Times

24-hour clock	12-hour clock	Some ways to say it
05.15	5.15 am	oh five fifteen five fifteen in the morning five fifteen am
10.30	10.30 am	ten thirty in the morning ten thirty am
14.45	2.45 pm	fourteen forty-five two forty-five in the afternoon two forty-five pm
21.55	9.55 pm	twenty-one fifty-five nine fifty-five pm nine fifty-five in the evening

Months

January, February, March, April, May, June, July, August, September, October, November, December

Days

Monday, Tuesday, Wednesday, Thursday, Friday, Saturday, Sunday

Saying years

- *1998 = nineteen ninety-eight*
- *2000 = two thousand*
- *2008 = two thousand and eight* (BrE); *two thousand eight* (AmE)

Writing dates

- *2011-06-14* (yyyy-mm-dd) – ISO 8601: an international standard
- *14/06/11* (dd/mm/yy) – commonly used in Europe
- *06/14/11* (mm/dd/yy) – commonly used in the US
- *14th June 2011*
- *14 June 2011*
- *June 14, 2011*
- *June 14th, 2011*

Saying dates

- *the fourteenth of June, two thousand and eleven* (BrE); *two thousand eleven* (AmE)
- *June the fourteenth, two thousand (and) eleven*

3 Symbols

General warnings and safety symbols

danger/warning/caution/hazard

Specific hazards

flammable   toxic/poison   high voltage

Safety equipment or help

emergency exit/ fire exit   fire alarm

fire extinguisher   hospital   first aid

emergency stop

Prohibitions

no entry   no exit   no smoking

Some electrical symbols

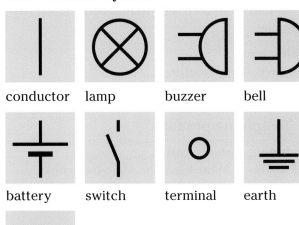

conductor | lamp | buzzer | bell
battery | switch | terminal | earth

fuse

Other symbols

+	plus/positive
–	minus/negative
#	hash/number
°	degree(s)
±	plus or minus
=	equals
≠	does not equal
≥	at least (also more than or equal to)
≤	up to (also less than or equal to)
~	approximately/about (also has other uses)
<	less than, under
>	more than, over
✓	tick
✗	cross
.	point (decimal number)

Currency symbols

€	euro(s)
$	dollar(s)/peso(s)/reai(s)
£	pound(s)
¥	yen
元	renminbi/yuan
﷼	rial(s)/riyal(s)
Rs Rp	rupee(s)

Internet symbols

@	at
.com	dot com
A-B	A hyphen B / A dash B
A/B	A slash B / A forward slash B
A_B	A underscore B

4 Useful words

Industries and technologies

aerospace
agriculture
automotive engineering
biotechnology
chemical engineering
civil engineering
building and construction
electrical engineering
electronics/electronic engineering
environmental engineering
information technology/IT
information and communications technology/ICT
manufacturing
marine engineering
materials testing
mechanical engineering
petroleum
public health
security
telecommunications/telecoms
transport

Names of jobs

engineer
manager
technologist
technician
supervisor
team leader
mechanic
operator

Materials

Metals: aluminium, titanium, copper, iron, lead, tin
Alloys: steel, chrome, cromoly
Plastics: polycarbonate, polyester, polystyrene, nylon
Composites: fibreglass, graphite

British and American English

Here are some of the words used in this book, but there are many more. Key the words *American British English* into an Internet search engine or *Wikipedia* to find complete lists. Some AmE words and spellings are now used also in BrE, for example, *antenna, disk*. Some BrE words are now used in AmE, for example, *car*.

British English (BrE)	American English (AmE)
accelerator	gas pedal/gas
aerial	antenna
aeroplane	airplane
aluminium	aluminum
cable/wire *(electricity)*	cord
car	automobile
centre	center
colour	color
disc	disk
earth *(electricity)*	ground
fibreglass	fiberglass
flat *(battery)*	dead
lift *(in a building)*	elevator
litre	liter
metre, kilometre, millimetre	meter, kilometer, millimeter
mobile/mobile phone	cellphone
petrol	gas/gasoline
polystyrene	styrofoam
postal code	zip code
spanner	wrench
storey *(in a building)*	floor/story
torch	flashlight
tyre	tire
vice *(in a workshop)*	vise
windscreen	windshield

5 Social phrases

Meeting a friend or co-worker
Hello. Hi. Morning. Good morning.
How are you? How are things? How are you doing? How's it going?
Fine, thanks. Great. How about you?

Introducing yourself
I'm Hans. My name's Hans.

Introducing someone else
This is Mia. She's a student here. She's a technician.

Meeting someone for the first time
Pleased to meet you. Nice to meet you. Good to meet you

Taking leave
Goodbye. Bye. Cheerio.
See you. See you later. See you tomorrow.

6 Telephone phrases

Beginning a phone call
Hello. This is Mike. It's Mike. Mike here. Mike speaking.
Hello. Is that Mike?
Yes, this is Mike. Is that Jim?

Listening to a voicemail
Thank you for calling ABC Computers.
You've reached the voicemail of John Wilson.
Please leave a message after the tone.

Leaving a voicemail
Hello. My name is …
My phone number is …
My email address is …
My address is …
I'd like to order/buy …
I'd like some information about …
Could you please send me your catalogue/brochure.
Please call me back. It's urgent.
Please get back to me when you can. Thanks.
Thank you.

Listening to an automatic message
Thank you for calling ABC Computers.
For the sales department, please press 1.
To hear information about our services, press 2.
To speak to a service technician, please hold. Please wait.

Answering a call from a customer
Thank you for calling ABC Computers.
This is the service department.
My name's Jason. This is Jason. Jason speaking.
I'm the service technician.
How can I help? How can I help you? What can I do for you? What's the problem?

7 Forms and email conventions

Forms

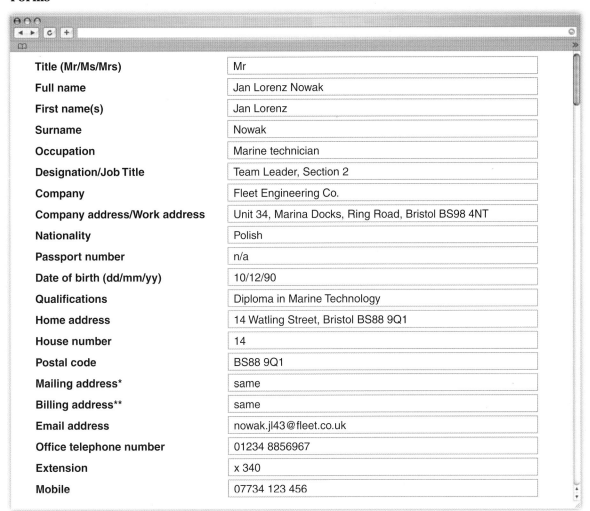

* we'll send the goods to this address
** we'll send the invoice to this address

Email

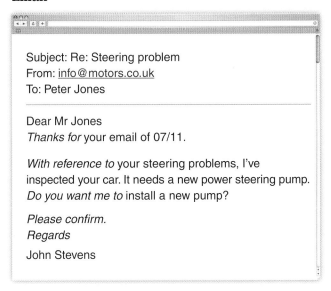

Hi Peter*
Thank you for
Re/With regard to/Concerning
Would you like me to
Please let me know. Best regards/Best/Best wishes John*

* Use this form when you know your customer well.

Extra material

2 Parts (1) 3 Ordering

Speaking exercise 4 page 14

Student A

1 Listen to Student B and make notes like this.

2 Change roles. Leave phone messages for Student B. Use the business cards below. Spell out the name of the person, and the company.

Example:
Hello. This is John West. That's W-E-S-T. Manager of Kesko. That's K-E-S-K-O. My phone number is 00 44 1224 867 4490. Please call me back.

Call from John West, Manager
Company: Kesko
Phone number: 00 44 1224 867 4490
Please call him back.

2 Parts (1) 3 Ordering

Task exercise 5 page 15

Student A

1 You are the sales person. Student B (the customer) telephones you. Ask Student B what they want to buy.

Item	Colour			Size			Quantity		
Helmet	red	(yellow)	blue	(large)	medium	small	1	(2)	3
Deck	red	yellow	(blue)	large	(medium)	small	1	2	3
Pad	(red)	yellow	green	large	medium	(small)	2	4	(6)

Useful phrases
What size/How many/What colour do you need?
What's your name? Please spell that. What's your phone number?

2 Change roles. You are a customer. You want to buy the items circled in blue. Telephone Student B (the sales person) and order the items.

Begin:
A: Hello. I need to buy some things for my skateboard.

3 Circle new items and phone up to order them.

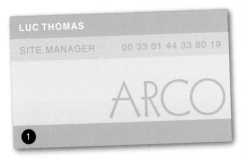

Extra material

Review A

Exercise 16 page 18

3 Parts (2) 3 Locations

Task exercise 8 page 25

Student A

1 Ask Student B where these items are and write them in their correct locations: *speakers, mouses, notebook computer, headphones, cables, computer monitors, DVD players.*

2 Then change roles. Answer Student B's questions.

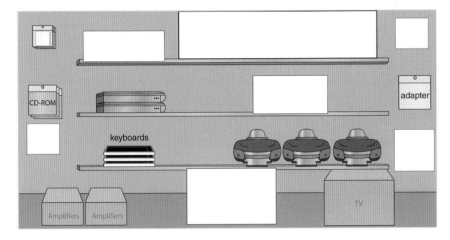

Here are some useful phrases:
- *on the top/middle/bottom shelf*
- *to the left/right of the shelves*
- *in/at the top/middle/bottom*
- *on the left/right*
- *above/below the shelves*

2 Parts (1) 1 Naming

Speaking exercise 11 page 11

Check your answers.

1 racing car	2 rocket	3 mountain bike
4 plane	5 motorbike	6 boat

Unit 12 Checking and confirming 1 Data

Speaking exercise 7 page 91

Student B

Confirm or correct Student A's answers.

Mars

1 6747 km
2 24 hours and 37 minutes
3 228 million km (average)
4 687 Earth days

Yes, that's right.
No, that's wrong. Change it to … .

Unit 12 Checking and confirming 2 Instructions

Speaking exercise 8 page 93

Write down what is happening in the pictures using the words in the box.

astronaut   car   helicopter
motorboat   plane   rover
shuttle   truck

Unit 11 Cause and effect
1 Pistons and valves

Start here exercise 1 page 84

Check your answers.

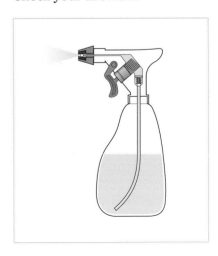

Extra material

5 Flow 1 Heating system

Task exercise 8 page 37

Student A

1. Explain one of these systems to Student B.
2. Listen to Student B, and ask questions. Then draw a simple diagram of his/her system.

Electric water heater

Gas water heater

6 Materials 3 Buying

Speaking exercise 3 page 46

Student A

Put together different components to make four email addresses and four web pages. Then dictate the addresses to your partner.

james.bond007	microsoft.com
roger.federer37	mozilla.com
leonardo.di.caprio89	toyota.co.fr
danielcraig19	citroen.com

(@ between columns)

www.	microsoft.com	/sales-department	/index.html
	mozilla.com	/service_and_repairs	/italian_pages
	toyota.co.fr	/catalogue.search	/new-products
	citroen.com	/new_ideas	/form-downloads.pdf

Example:
A: What's your email address?
B: It's danielcraig19@mozilla.com.
A: (Writes it down.) Do you have a website?
A: Yes, I do.
B: What's the web address?
A: It's www.mozilla.com/new_ideas
B: (Writes it down.) Thanks.

7 Specifications
1 Dimensions

Task exercise 9 page 53

Student A

2. Answer Student B's questions about the Rion-Antirion Bridge.

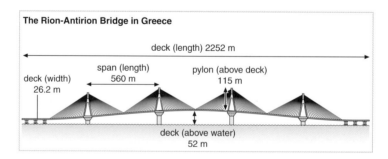

8 Reporting
2 Damage and loss

Task exercise 9 page 61

Student A

2   Answer Student B's questions about the damage to your car.

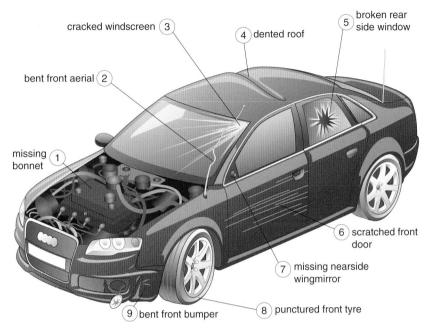

10 Safety
3 Investigations

Task exercise 5 page 79

Student A

1   Read about your incident and answer Student B's questions.

> Two days ago, 23rd November, a builder called Gino Petri had an accident on the 3rd floor of the new building. The accident happened at 09.38. Mr Petri was about 20 m above the ground at the time. He tripped over a metal girder and he fell from the 3rd floor to the 2nd floor. He fell into a safety net and received no injuries from the fall, but the girder cut his leg.

2   Then change roles. Investigate Student B's incident. Ask questions and complete the report form on page 79.

4 Movement   3 Actions

Task exercise 7 page 31

Student A

1   You're learning to drive the truck. Student B is your driving instructor. Follow Student B's instructions and rearrange your pictures into the correct sequence.

The correct sequence of the instructions is:

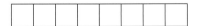

2   Then change roles. Tell Student B to follow these instructions in the correct sequence.

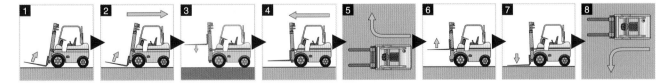

6 Materials 3 Buying

Task exercise 6 page 47

Student A

1. You are the sales person in the sports shop. Ask Student B questions and complete this order form. Ask about the features they want (size, colour, material), and the price.

2. Then change roles. You are now the customer. Circle three items you would like to buy, and circle the features you want and the price. Then phone up the shop and place your order. You can either make up details (e.g. names, phone numbers, etc.) or use your own.

USEFUL LANGUAGE

- What's your name/phone number/ email address?
- Could you spell/repeat that, please? Is that six<u>teen</u> or <u>six</u>ty?
- What's the product name/number?
- What colour/size/material would you like/do you need?
- Do you want to pay in dollars ($), sterling (£) or euros (€)?
- How many would you like/do you need?

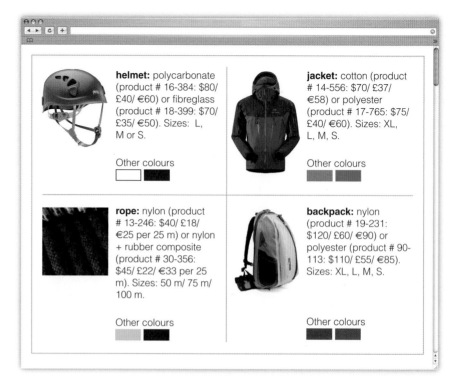

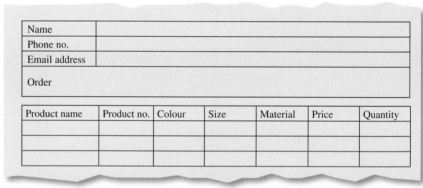

8 Reporting
2 Damage and loss

Task exercise 9 page 61

Student B

1. Answer Student A's questions about the damage to your car.

2. Then change roles. Now ask Student A questions about the damage to their car. Turn back to page 61. Label your diagram.

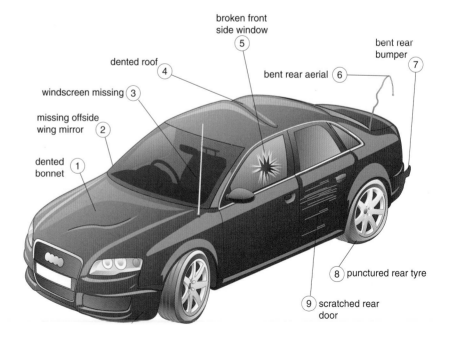

Extra material

9 Troubleshooting
2 Hotline

Task exercise 8 page 71

Student A

Find out all the differences between your wiring diagram and your partner's.

Hint: there are at least ten differences of (a) location of sockets and (b) wiring connection.

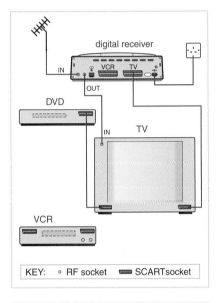

KEY: ○ RF socket   ▭ SCART socket

USEFUL LANGUAGE

digital receiver, DVD, VCR, TV, antenna, SCART socket, RF socket, in, out, power, socket

Do you have a/an ... ?
Look at the
Where is the ... ?
Does the ... connect to the ... ?
Have you connected the ... to the ... ?
Is the ... connected to the ... ?

Unit 12 Checking and confirming 3 Progress

Task exercise 5 page 95

Student A

It's 8th August. Answer Student B's questions about your chart.

Task	August											
	2	3	4	5	6	7	8	9	10	11	12	13
Dismantle old water system	■	■	■									
Assemble new water system				■	■							
Install water system						■	■					
Test equipment for third spacewalk					■							
Take video of damaged nose cap						■						
Inspect damage to waste tank							■	■				
Assemble new robot arm									■			
Attach new robot arm										■	■	■

B: Have you dismantled the old water system yet?
A: Yes, we have.
B: When did you complete the job?

Unit 8 Reporting 1 Recent incidents

Speaking exercise 7 page 59

Look at this picture for one minute. Then turn back to page 59.

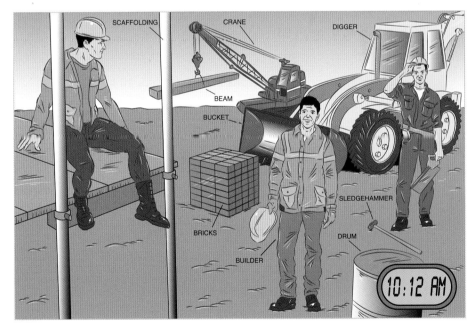

Extra material   117

6 Materials 3 Buying

Speaking exercise 3 page 46

Student B

Put together different components to make four email addresses and four web pages. Then dictate the addresses to your partner.

william.tell.17		apple.co.uk
david.bowie88	@	siemens.co.de
sean.penn519		UPS.com
michael.schumacher		vodafone.com

www.	vodafone.com	/country/uk/en	/manual/download.doc
	siemens.co.de	/technical_support	/welcome.html
	UPS.com	/country/us/en	/support.html
	apple.co.uk	/training-and-development	/application_form.pdf

Example:
A: What's your email address?
B: It's william.tell.17@apple.co.UK
A: (Writes it down.) Do you have a website?
A: Yes, I do.
B: What's the web address?
A: It's www.apple.co.uk/country/uk/en
B: (Writes it down.) Thanks.

7 Specifications 1 Dimensions

Task exercise 9 page 53

Student B

1 Answer Student A's questions about the Akashi-Kaikyo Bridge.

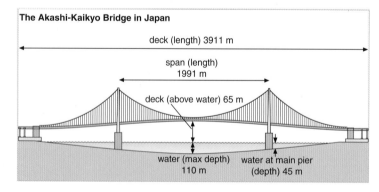

The Akashi-Kaikyo Bridge in Japan
deck (length) 3911 m
span (length) 1991 m
deck (above water) 65 m
water (max depth) 110 m
water at main pier (depth) 45 m

2 Then change roles. Ask Student A questions about the Rion-Antirion bridge. Complete your specification chart.

10 Safety
3 Investigations

Task exercise 5 page 79

Student B

2 Read about your incident and answer Student B's questions.

Yesterday, 15th July, an electrician called Pedro Gomez had an accident on the #1 scaffolding. The accident happened at 14.46. Mr Gomez was about 10 m above the ground at the time. He raised his right arm. His arm touched a live wire and received a small electric shock. He had a small 2 cm burn on his right arm, but received no other injuries.

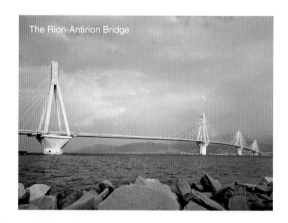

The Rion-Antirion Bridge

Rion-Antirion Bridge: specifications	
Type of structure	Cable-stayed
Country	
Piers (number)	
Span (length)	
Deck (above water)	
Deck (length)	
Deck (width)	
Pylon (above deck)	

Audio script

Unit 1 Check-up

🔊 02

1. A: Hello. I am Hans Beck.
   B: Hi. My name is Pedro Lopez.
   A: Pleased to meet you.

2. A: Excuse me. Are you Mr Rossi?
   B: Yes, I am.
   A: Pleased to meet you, Mr Rossi. I'm Danielle Martin.
   B: Nice to meet you, Danielle.

3. A: Hi. My name's Jamal.
   B: Hello, Jamal. I'm Borys.
   A: Good to meet you, Borys. Are you from Russia?
   B: No, I'm from Poland.

🔊 03

OK, please follow these instructions.
Please stand up.
Sit down, please.
Stand up again.
Please stand up again.
Raise your left arm.
Please raise your left arm.
Lower your arm, please.
Now raise your right arm.
Please raise it.
Now lower your arm, please.
OK, sit down.
Sit down!
Sit down, please.
Write your name, please.
Now say your name.
Please say your name.
Say Hello.
Say Hello, please.
Please pick up a book.
Please read it silently.
Now read it aloud.
Read it aloud, please.
Stop!
Stop!
Please stop.
Please be quiet.
Please say Goodbye.

🔊 04

I'm Bruno Martyn. That's M-A-R-T-Y-N. My phone number is oh oh three three, oh five six two, one nine, eight five, six four. My email address is mart seventeen at macrosoft dot co dot fr, that's M-A-R-T seventeen at macrosoft dot co dot fr.

🔊 05

1

[R = Receptionist; Q = Mr Quayle]

R: Welcome, sir. Could you give me your surname, please.
Q: Yes, it's Quayle. Q-U-A-Y-L-E.
R: And your company name, sir?
Q: It's Vox.
R: How do you spell that?
Q: V-O-X
R: Thank you. And your email address, sir?
Q: It's pq99 at biz.com. That's P-Q ninety-nine at biz.com. That's B-I-Z dot com.

2

[PO = Phone operator; M = Ms Mathers]

PO: Emergency, which service?
M: Fire.
PO: Right, what's your address?
M: 17 East Street.
PO: Repeat the address, please.
M: 17 East Street.
PO: How do you spell East?
M: E-A-S-T.
PO: What's your postcode?
M: CS4 8NT.
PO: Repeat your postcode, please.
M: CS4 8NT.
PO: And your surname, please.
M: Mathers.
PO: How do you spell that?
M: M-A-T-H-E-R-S.
PO: Thank you.
→

3

[CS = Customer Services; PB = Pieter Braun]

CS: This is Customer Services. How can I help you?
PB: My radio doesn't work.
CS: Oh, I'm sorry to hear that, sir. All right, please give me some details. What's your full name?
PB: Pieter Braun.
CS: How do you spell your surname?
PB: B-R-A-U-N.
CS: Thank you, Mr Braun. And what's your postcode?
PB: 20953.
CS: Thank you, and your house number please?
PB: 67.
CS: Thank you, sir. And what's the model number of the radio?
PB: GJ 8041.
CS: Could you repeat that, please?
PB: GJ 8041.
CS: Thank you.

06

1 Counter number 11, please.
2 This is Radio 1 on 98.8 FM.
3 Please pay 18 pounds and 80 pence.
4 The 14.43 train to Oxford will depart from platform number 9.
5 Flight number EZ 370 is boarding now. Please go to gate number 14.
6 To donate money to Live Aid, ring this number now: 0207 903 8672.
7 Begin countdown now: 20, 19, 18, 17, 16, 15, 14, 13 …

07

eighty euros
fifteen degrees
thirteen amps
eighty-nine degrees Celsius
forty watts
one point two kilometres
thirteen point eight metres
one hundred and ninety kilometres per hour
one hundred and fifty thousand litres
twelve thousand five hundred revolutions per minute
two hundred and thirty volts
one hundred and sixty kilograms

08

Here are the results of the finals of the men's 1500 metre race:

In first place, it's El Guerrouj from Morocco. His time is three minutes, thirty-four point one eight seconds.

In second place, it's Lagat from Kenya. His time is three minutes, thirty-four point three oh seconds.

In third place, it's Silva from Portugal. His time is three minutes, thirty-four point six eight.

In fourth place, it's Timothy Kiptanui from Kenya. His time is three minutes, thirty-five point six one.

In fifth place, it's Heshko from the Ukraine. His time is three minutes, thirty-five point eight two.

In sixth place, it's Mike East from Britain. His time is three minutes, thirty-six point three three.

09

1 The 28th of December 2010.
2 The 18th of November 2008.
3 The 21st of July 1999.
4 The 12th of January 2009.

10

1 LH 306 departs from Frankfurt at seven thirty am and arrives in Warsaw at nine oh five am.
2 AF 835 departs from Paris at eight twenty am and arrives in Madrid at ten ten am.
3 EK 971 departs from London at six thirty am and arrives in Bahrain at three fifteen pm.
4 MS 740 departs from Dubai at two forty pm and arrives in Cairo at five fifty pm.
5 AZ 7788 departs from Rome at nine ten pm and arrives in Tokyo at four fifteen pm the next day.
6 SA 104 departs from Johannesburg at three forty-five pm and arrives in Lagos at nine twenty-five pm.

11

1 It's eighteen thirty-five on the fifteenth of September.
2 It's eight fifty-five on the fifth of November.
3 It's thirteen forty-five on the thirteenth of December.
4 It's fourteen fifty-five on the thirtieth of October.

Unit 2 Parts (1)

12

The world record for a high jump on a skateboard is 7.1 metres. A young skateboarder, Danny Way, jumps 7.1 metres on the 19th of June 2003.

The world record for a long jump on a skateboard is 24 metres. Skateboarder Danny Way jumps 24 metres on the 8th of August 2004.

13

1 tail **2** truck **3** deck **4** nose
5 wheel **6** axle **7** plate

14

A: What's this called?
B: It's called a deck.
A: What's this called in English?
B: It's called a truck.

16

[C = Customer; S = Shopkeeper]

C: Hello.
S: Good morning. What can I do for you?
C: I need a spanner, please.
S: What size do you need?
C: Erm, I think it's ten millimetres.
S: OK. Here you are. One ten millimetre spanner.
C: Thanks. And I need some nuts, please.
S: Some nuts, did you say? OK, what size do you need?
C: Erm … seven mil.
S: Right. And how many do you need?
C: Four.
S: Right. Here you are. Anything else?
C: Yes, I need some bolts, please.
S: What size?
C: M5.
S: And how many M5 bolts do you think you need?
C: Eight, please.
S: OK, here you are.
C: Thanks.

18

Thank you for calling Skateboards 4 U. Please leave a message after the tone.

Erm, Hello. Erm, I need some parts … er … for my skateboard. My name is Ben, Ben Johnson. That's J-O-H-N-S-O-N. My er … my phone number is … double oh, double 4, 208 8947. Please call me back. Thanks.

19

1 Abdul … that's A-B-D-U-L Monim … spelt M-O-N-I-M Waheed … that's W-A-H-E-E-D, and my phone number is 00 202 48830.

2 José … that's spelt J-O-S-E Fernando … that's F-E-R-N-A-N-D-O Ruiz … that's R-U-I-Z. Phone number 00 35 912 828 990.

3 Adil spelt A-D-I-L Al-Mansur … that's A-L hyphen M-A-N-S-U-R. Phone number 00 971 2 605 9943.

4 Nikolai that's N-I-K-O-L-A-I Kuznetsev … that's spelt K-U-Z-N-E-T-S-E-V. Phone number 00 7 455 988-22-77.

20

A: I'm Luis. I'm a student. And this is Paulo. He's a student, too.
B: Hello, Luis. Hello, Paulo. Nice to meet you.

Unit 3 Parts (2)

21

This is the new Multi Tool! Use it at home. Use it on the building site. Use it when you travel. It has a hammer and a pair of pliers. It also has a saw, a blade and a can opener. The Multi Tool has everything you need! Only £29.99. Buy one now!

22

A: Do you have a Multi Tool?
B: Yes, I do.
A: Does the Multi Tool have a hammer?
B: Yes, it does.
A: Does it have a pair of scissors?
B: No, it doesn't.

25

A: OK, now put the cursor on the START button.
B: Where's the START button?
A: It's at the bottom. On the left. Do you see it?
B: Yes. Is that it?
A: Yes, that's correct. … Now, move the cursor up to the CLOSE button.
B: Where's that?
A: It's an X. It's on the right. At the top.
B: Is that it?
A: Yes, that's it. Now click.

Unit 4 Movement

26

1 19 degrees.   2 40 degrees.
3 70 degrees.   4 118 degrees.

27

1 A fast CD-ROM can rotate at 9800 revolutions per minute.
2 Sound travels at about 1200 kilometres per hour.
3 The maximum land speed is about 1228 kilometres per hour.
4 The maximum speed of a boat on water is about 154 metres per second.
5 The Earth rotates at 1000 miles per hour.
6 The Earth moves around the Sun at 67,000 miles per hour.

Unit 5 Flow

28

[L = Lecturer; S = Student]

L: Right. Now let's look at this diagram of the circuit, up here. Can you see it clearly? On the left, here, you can see a solar panel. OK? The solar panel collects the sunlight and changes it into electricity. And here, on the right, you can see three lamps. These three long things. OK? And there, between the panel and the lamps, you can see a controller and a battery.
S: Excuse me, sir. Which one is the controller?
L: Well, the controller's at the top, OK? And of course the battery's at the bottom, here, below the controller. And finally, you can see some electrical cables or wires. The cables run from the panel, through the controller, into the battery, and also into the lamps.

29

[L = Lecturer; S = Student]

L: OK? So to summarise, here again, these are the main parts of the system. A sixty watt solar panel; ... a five amp controller; ... a twelve volt one hundred ampere hours battery; ... and three twelve volt eight watt lamps.
S: Excuse me, what kind of electrical current is it?
L: It's a direct current – DC. Is that clear now?

30

1 A: Turn down that thermostat, please. The water's too hot for a shower. The correct temperature is about 60 degrees.
   B: Fahrenheit?
   A: No, Celsius, of course.
2 A: This refrigerator is too cold. Turn the temperature up to about 4.5 degrees.
   B: Fahrenheit?
   A: No. That's too cold. 4.5 degrees Celsius.
3 A: That freezer's too warm. Turn the temperature down to zero degrees.
   B: Zero degrees Celsius?
   A: No, that's too warm. Zero degrees Fahrenheit. That's the same as minus eighteen degrees Celsius.
4 A: Do you know the record for the coldest air temperature in the world?
   B: No.
   A: It's minus 89 degrees.
   B: Fahrenheit?
   A: No, Celsius.
   B: Where is it?
   A: In Antarctica.
5 A: And the hottest temperature in the world. Do you know that?
   B: No.
   A: It's 136 degrees.
   B: Celsius?
   A: No, no. Fahrenheit.
   B: Where is it?
   A: In Libya.
6 A: The car engine is too hot.
   B: Why? What's the correct temperature?
   A: About 110 degrees.
   B: Is that Fahrenheit?
   A: No, Celsius.

31

[D = Dan; J = Jack]

D: I work in the electronics workshop every Thursday and Friday.
J: When do you attend lectures?
D: Every Tuesday morning.
J: What do you do on Tuesday afternoons?
D: I do my practical work then.

Unit 6 Materials

32

[L = Lecturer; T = Trainees]

L: Today, we're doing a tensile strength test for this mountaineering rope. OK. Is everyone ready? Can you see and hear me clearly?
T: Yes.
L: All right, now listen and watch carefully. The rope is made of nylon. Now I'm pulling the rope. I'm stretching it. Is it breaking?
T: No.
L: That's right. It isn't breaking.

35

Hello, This is Manuel Vargas. That's V-A-R-G-A-S. My phone number is double oh, 34 94 double 6 389. I'll repeat that: double oh, 34 94 double 6 389. Please send me your catalogue of sports equipment. My email address is mvargas17@xtreme_sports.co.es. I'll say that again, mvargas17 that's M-V-A-R-G-A-S seventeen all one word … at … xtreme underscore sports, that's spelt X-T-R-E-M-E underscore S-P-O-R-T-S dot co dot E-S.

36

1 waleed at sports dot com
2 adam at city dot co dot U, K
3 theo walcott, that's T-H-E-O then W-A-L-C-O-T-T at goalfeast, that's G-O-A-L-F-E-A-S-T, all one word dot com
4 C dot ronaldo, that's R-O-N-A-L-D-O at back-of-the-net, that's B-A-C-K dash O-F dash T-H-E dot net
5 WWW dot toyota, that's T-O-Y-O-T-A dot com forward slash customer dash support
6 WWW dot orascom, that's O-R-A-S-C-O-M dot com dot E-G forward slash sales underscore one

37

A: What's your surname, please?
B: It's Lint.
A: Could you repeat that, please?
B: Lint.
A: Could you spell that, please?
B: L-I-N-T
A: Is that T or D?
B: It's T. T for teacher.
A: Thanks. And what's the product number?
B: It's seventeen dash three oh five.
A: Is that 17 or 70?
B: Teen. Seventeen. One seven.
A: Right. Thanks.

38

[J = John, M = Mike]

Dialogue 1
J: Hello?
M: Hello. Is that John?
J: Yes?
M: It's Mike.
J: Oh hi, Mike.
M: Hi. How are you?
J: OK, thanks. How are you?
M: Fine. I want to ask you …

Dialogue 2
J: Hello?
M: Hello. Is that John?
J: Yes. Is that Mike?
M: Yes, it's me. Hi. How are you?
J: Fine, thanks. How about you?
M: I'm fine. Would you like to …

Dialogue 3
J: Hello. John Davis here.
M: Oh hi, John. This is Mike.
J: Hi, Mike.
M: Hi. How are things?
J: Great, thanks. How are you?
M: Good. I'm phoning you to …

Unit 7 Specifications

39

This is a photograph of the Millau road bridge. That's Millau, spelt M-I-L-L-A-U. It's a beautiful bridge and it's very high. In fact, it's one of the highest bridges in the world. It's in the south of France and it crosses the river Tarn. The bridge is three hundred and thirty-six point four metres above the river.

40

[P = TV presenter; E = Engineer]

P: Yes, the total height of the Millau road bridge is 336.4 metres above the river Tarn. Now I'm talking to the chief engineer of the bridge. So, can I check with you? Three hundred and thirty-six point four is the total height from the top of the pylons to the river, is that right?

E: Yes, that's right. That's the total height. The road deck itself is 246 metres above the river. Then the pylons, or towers, rise another 90 metres above the deck.

P: I see. And how wide is the river valley at the bridge?

E: Well, the valley is almost 2.5 kilometres wide. In fact, the bridge is 2460 metres long, or 2.46 kilometres.

P: And how long are the spans?

E: They have different lengths. The bridge has two outer spans and six inner spans. The two outer spans are 204 metres long, and the six inner spans are 342 metres long.

P: How wide is the road deck?

E: It's 32 metres wide. It has four lanes of traffic.

P: And what's the bridge made of? It's really beautiful and it looks very light.

E: Yes, it looks light because it is light. It uses the minimum material. But it's also very strong. The cables and the road deck are in fact made of steel. The seven piers, of course, are made of reinforced concrete.

41

Picture 1 is Taipei 101 in Taiwan. Its height is 508 metres.

Picture 2 is the Shanghai World Financial Centre in China. Its height is 492 metres.

Picture 3 is the Abraj Al Bait Towers in Saudi Arabia. Its height is 485 metres.

Picture 4 is the Petronas Towers in Malaysia. Its height is 452 metres.

Picture 5 is the Federation Tower in Russia. Its height is 448.2 metres.

Picture 6 is the Dubai Towers in Doha, Qatar. Its height is 445 metres.

Picture 7 is the Sears Tower, in the USA. Its height is 442 metres.

42

[T = Tom; Dr J = Dr Jensen]

T: Today on RadioTech, I'm talking to Dr Tore Jensen. He's a civil engineer and his company is working on plans for a tunnel under the Atlantic Ocean. So, Tore, tell me about this tunnel, or tube, under the Atlantic. Are you building it now?

Dr J: No, no, we're not building it now. That's a long time in the future. Right now, we're thinking about it and planning it. Another company is designing a small-scale model.

T: So, when will they build it?

Dr J: I think they'll start in 2080 and complete it in 2100.

T: Wow! That is a long time in the future.

Dr J: Yes, it is!

T: So, where will the tunnel be? How long will it be? How deep?

Dr J: The tube will be below the Atlantic Ocean. It'll connect the USA with Britain. It'll be about 5000 km long and about 100 metres deep in the ocean.

T: Will the tube move around in the water?

Dr J: No, it won't move. One hundred thousand cables will attach it to the sea floor.

T: Will the train use electricity?

Dr J: No, it won't. It'll use magnetism. The tube will contain a vacuum. MagLev trains will be able to travel through the tube at 8000 km/h.

Unit 8 Reporting

43

[PO = Phone operator; D = Driver]

PO: Hello. Crash Recovery Company. How can I help you?

D: Oh hi! I've broken down on the motorway!

PO: OK, don't worry. What's your name and car registration number?

D: My name's Anita Zubaid. That's Anita, spelt A-N-I-T-A Zubaid, spelt Z-U-B-A-I-D. The car is Y449 MNE.

PO: And where are you, Ms Zubaid?

D: I'm on the M13. Between Junctions 15 and 16. Going south.

PO: Right. And what's the problem?

D: Well, the exhaust pipe has fallen off.

PO: OK. We'll be there in 30 minutes. Stay with your car, please.

D: All right. Bye.

44

Call number 1

[C1 = Caller 1; S = Security]

C1: Hello? Hello? Is that Security?
S: Yes, Security here. How can we help?
C1: Some thieves have broken into my office. They've taken my computer.

Call number 2

[IT = IT technician; C2 = Caller 2]

IT: IT department. How can I help you?
C2: Is that the IT hotline?
IT: Yes. What's the problem?
C2: Something has happened to my computer. I've lost all my data.

Call number 3

[EO = Emergency phone operator; C3 = Caller 3]

EO: Emergency. Which service?
C3: I need an ambulance, quickly.
EO: What's happened?
C3: It's my daughter. She's fallen down some stairs. She's cut her leg.

Call number 4

[PO = Phone operator; C4 = Caller 4]

PO: Crash Recovery. How can I help you?
C4: Oh, hello. Yes. I've had an accident. I've driven my car into a bridge.

45

[D = Del; Mr E = Mr Ericsson]

D: Customer Services. Del speaking. Please give me your order number.
Mr E: AX 5831-77 …
D: Ah yes, Mr Ericsson. You've bought a radio from us. How can I help you?
Mr E: I've opened the box and taken out the radio. There's some damage and there are some missing items.
D: I'm sorry to hear that. What's missing?
Mr E: The power cable has no plug. …
D: No … plug … on … cable. OK. Anything else?
Mr E: Yes. There are no batteries and no headphones. …
D: No batteries … and no headphones. OK. Is that all?
Mr E: No. There are no cables for the speakers and there's no user manual. …
D: Cables for speakers and user manual … missing. Anything else?
Mr E: There's some damage. The body is cracked. There's a scratch on the screen. …
D: Screen … scratched. Body … cracked. OK. Is there any more damage?
Mr E: Yes. The antenna is bent and the speakers are dented. And there are some holes in one speaker. …
D: Oh dear, I do apologise for all that. Please put everything in the box again. We'll collect it from your house tomorrow. Then we'll send you a new radio.
Mr E: OK.
D: Goodbye, sir. Thanks for calling.

46

[CS = Customer Services; BJ = Ben Jones]

CS: Hello, Electronic Repairs. Don speaking. How can I help you?
BJ: Hi. My name's Ben Jones. I've broken my MP3 player. Can you repair it?
CS: OK, sir. What's the model number?
BJ: It's a Super 30 GB.
CS: And when did you buy it?
BJ: Er, let's see … . Yes, I bought it on the 18th of August.
CS: And what's the problem?
BJ: I've dropped it and I've cracked the screen.
CS: And, er … when did you crack the screen?
BJ: Yesterday.
CS: OK, bring it into the shop and I'll look at it.
BJ: Thanks. Bye.

Unit 9 Troubleshooting

47

A: Look at the airboard. You can see the five main parts: the body, the engine, the fan, the handlebar and the two levers. The body supports the rider, and the engine drives the fan. The handlebar steers the airboard left and right.
B: Ah yes, I see. So what does the fan do?
A: It pulls the air in and forces it downwards.
B: Right. And what do the two levers do?
A: They control the speed and acceleration of the airboard.

48

Thank you for calling New Tech. For the sales department, press 1. For the service department, press 2.

This is the service department. For computers, press 3. For printers, press 4.

This is the computer unit. To hear information about our services, press 5. To speak to a service technician, press 6.

49

Hello, you've reached the computer service hotline. This is Jan speaking. I'm the technician. How can I help you?

50

[C = Customer; ST = Service technician]

C: Hello, is that the IT hotline?
ST: Yes, it is. I'm the technician. My name's Sofia. How can I help you?
C: I've got a problem with my wireless router. It doesn't work.
ST: OK. I'll talk you through it. Are you sitting at the computer now?
C: Yes, I am.
ST: OK. Look at the back. Is the router connected to the power outlet?
C: Yes, it is.
ST: OK. And is the router connected to the modem? That's the green cable.
C: Ah … no, it isn't.
ST: So, connect the router to the modem now. … Have you done that?
C: Yes, I have. I've connected it.
ST: OK. Now, have you connected your computer to the router? That's the blue cable.
C: Erm … no, I haven't.
ST: OK. Do it now. … Have you done that?
C: Yes, I have.
ST: OK. Now let's look at the lights …

51

1 A: Are the lights on?
   B: Yes, they are.
2 A: Is the computer connected to the adapter?
   B: No, it isn't.
3 A: Have you sent the email?
   B: Yes, I have.
4 A: Does your new radio work?
   B: No, it doesn't.
5 A: Did you go to the cinema yesterday?
   B: No, I didn't.
6 A: Can I speak to your brother?
   B: Yes, you can.
7 A: Do you work in the city?
   B: Yes, I do.
8 A: Are you sitting at the computer now?
   B: No, I'm not.
9 A: Do those speakers cost a lot of money?
   B: No, they don't.
10 A: Has your car broken down?
   B: Yes, it has.

52

A: Press the power button.
B: OK. I'm pressing it.
A: Does the computer start?
B: No, it doesn't.
A: OK. Check the green LED.

Unit 10 Safety

53

1 You must wear a hard hat on the building site.
2 Don't go through that door!
3 You must wear safety gloves everywhere in the factory.
4 Don't touch that machine! It's very hot.
5 Be careful! High-voltage electricity. You might get an electric shock.
6 You mustn't use your mobile phone here.

54

1 Look out! There's a low beam in front of you.
2 Be careful! There are some bricks on the floor.
3 Watch out! There's no guard on the gears.
4 Mind the gap! There's a gap between the train and the platform.
5 Careful! There are bare electrical wires on the wall.
6 Look out! The water is very hot.

55

[AC = Air traffic controller; P = Pilot]

AC: ConAir 286. Unknown traffic. Two o'clock. 150 metres. Crossing right to left.
P: ConAir 286. Negative contact. Request vectors.

AC: Turn right. Heading 045. Descend. 85 metres.
P: Right turn. Heading 045. Descending. 85 metres. ConAir 286. …
AC: ConAir 286. All clear. Resume own navigation.
P: Roger. ConAir 286.

Unit 11 Cause and effect

56

1 *[Urgent sound of alarm bell]*
2 *[Sound of beep in automatic phone]*
3 *[Sound of buzzer]*
4 *[Sound of car horn]*
5 *[Sound of dial tone after picking up phone]*
6 *[Sound of door bell, ding-dong]*
7 *[Sound of mouse click]*
8 *[Sound of siren]*

57

The German company Enercon manufactures the world's tallest wind turbine. The tower of this huge turbine, the Enercon E112, is 186 metres tall. But the world's highest wind turbine is about 2300 metres up a mountain in Gütsch in Switzerland. The tower of the wind turbine isn't very tall, but at 2322 metres, it's the highest in the world.

Wind turbines start producing power at the minimum wind speed of about 15 kilometres per hour. If the wind speed is less than 15 kilometres per hour, the wind turbine doesn't switch on. The maximum wind speed for a turbine is about 90 kilometres per hour. If the speed of the wind is more than this, the turbine switches off and the blades stop.

Unit 12 Checking and confirming

58

[C = Controller; R = Rover]

C: Move forwards 200 cm.
R: Confirmed. I'm moving forwards 200 cm.
C: Now rotate 15 degrees to the left.
R: Confirmed. I'm rotating 15 degrees to the left.

59

[T = Trainer; Tr = Trainee]

T: Right. I'll give you an instruction. First, do it. Then confirm what you're doing, OK?
Tr: OK.
T: Then confirm what the rover's doing. Is that clear?
Tr: Yes.
T: Right. Let's go. First, make the rover move forwards 200 cm.
Tr: OK. I'm pushing the joystick forwards.
T: Good. Now what's happening?
Tr: The rover isn't moving.
T: Right. Wait five seconds. Now what's happening?
Tr: OK. It's moving forwards now.

60

We sometimes have to make many spacewalks outside the space station, just to do one simple repair job. Let me give you an example. A small piece of rock from space has hit an oxygen tank. What do we do?
First, we must test our equipment for the spacewalks.
Then, in the first spacewalk, we inspect the damage. We take photographs of the tank and the hole.
After that, we go back into the space station. There we plan the repair and prepare for the next spacewalk.
In the second spacewalk, we disconnect the pipes from the tank – these pipes carry the oxygen into the space station. We remove the tank. Then we bring the tank into the space station.
Back in the space station, we dismantle the tank. We repair the damage. If this isn't possible, we replace the part.
Then we assemble the tank again.
In the third spacewalk, we attach the tank to the side of the space station and connect the pipes to the tank.

61

[C = Controller; A = Astronaut]

C: OK, today is the 6th of June, 7 pm in the evening. I'm checking progress on the space station. Have you done the first spacewalk yet?
A: Yes, we have.
C: Good. When did you do it?
A: We did the spacewalk yesterday, on the 5th of June.
C: Right. And have you repaired the oxygen tank yet?
A: No, we haven't repaired it yet. We're still repairing it.
C: When will you finish it?
A: We'll complete the job tomorrow morning.

Pearson Education Limited

Edinburgh Gate
Harlow
Essex CM20 2JE
England

and Associated Companies throughout the world.

www.pearsonELT.com

© Pearson Education Limited 2008

The right of David Bonamy to be identified as author of this Work has been asserted by him in accordance with the Copyright, Designs and Patents Act 1988.

All rights reserved; no part of this publication may be reproduced, stored in a retrieval system, or transmitted in any form or by any means, electronic, mechanical, photocopying, recording, or otherwise without the prior written permission of the Publishers

First published 2008
Tenth impression 2014

ISBN 978-1-4058-4545-8

Set in Adobe Type Library fonts

Printed in China
GCC/10

Acknowledgements

We would like to dedicate this book to the memory of David Riley, whose tireless professionalism contributed so much to its creation and success.

The publishers and author would like to thank the following for their invaluable feedback, comments and suggestions, all of which played an important part in the development of the course: Eleanor Kenny (College of the North Atlantic, Qatar), Julian Collinson, Daniel Zeytoun Millie and Terry Sutcliffe (all from the Higher Colleges of Technology, UAE), Dr Saleh Al-Busaidi (Sultan Qaboos University, Oman), Francis McNeice, (IFOROP, France), Michaela Müller (Germany), Małgorzata Ossowska-Neumann (Gdynia Maritime University, Poland), Gordon Kite (British Council, Italy), Wolfgang Ridder (VHS der Stadt Bielefeld, Germany), Stella Jehanno (Centre d'Etude des Langues/ Centre de Formation Supérieure d'Apprentis, Chambre de Commerce et d'Industrie de l'Indre, France) and Nick Jones (Germany).

The author would like to thank Stephen Nicholl (Publisher) for his enthusiasm and dedication to the project, and for tempering his professional rigour with understanding and humour. He would also like to thank Eddi Edwards (Design Manager), Keith Shaw (Designer), Celia Bingham (Editor), Ben Greshon (Senior Editor), Kevin Brown (Picture Researcher) and Ann Oakley. Thanks also to Ian Wood for his early advice and support, and to Bruce Neale and Kate Goldrick.

The author would also like to thank his colleagues past and present around the world, the dedicated teachers of English, communication skills, science, technology, engineering, business and technical/ vocational skills, along with managers, supervisors, technicians and support staff, too many to mention, who have contributed more than they know by generously sharing their ideas and expertise. He would also like to say a special *dhanyavad* to his family and friends for their patience and unwavering support.

Illustrated by Mark Duffin, Peter Harper and HL Studios

The publisher would like to thank the following for their kind permission to reproduce their photographs:

(Key: b-bottom; c-centre; l-left; r-right; t-top)

Alamy Images: FAN Travel Stock 88; Royan Ong 28; Transtock Inc 11 (boat); alveyandtowers.com: 7 (E), 7 (G), 59; Art Directors and TRIP photo Library: 24, 44 (8), 47bl, 55, 116bl; aviation-images.com: 11 (Plane), 44 (plane); BAA Aviation Photo Library: 7 (C); Bigstone Ltd: 47tl, 116tr; Buzz Pictures: 10; Camera Press Ltd: 115; Construction Photography: 44 (beams); Corbis: 11 (rocket), 74; David Bebber / Reuters 26; EPA 62l; Lester Lefkowitz 6b; Michael Kim 45; Murat Taner 53; DK Images: 44 (racket), 44 (frame); Eye Ubiquitous / Hutchison: 80; Getty Images: 11 (racing car); David Lees 15; Jeff Haynes / AFP 8; iStockphoto: James Kingman 70t; Kaito Electronics Inc, www.kaitousa.com: 22t, 22b; Los Alamos National Laboratory: J L Lacour / CEA 90; Lyon Equipment Ltd: 47tr, 47br, 116tl, 116br; Martyn Chillmaid Photographer: Martyn Chillmaid / photographersdirect.com 52; Masterfile UK Ltd: 70b; NASA: 62c, 62r, 94; PA Photos: 78; Photolibrary.com: Brian Milne 58; PunchStock: Digital Vision 6t, 77; Image Source 11 (mountain bike), 12; Medio Images 44 (shoe); Purestock 44 (sunglasses); Stockbyte 4; Uppercut 44 (surf boards); Redferns Music Picture Library: Mike Cameron 7 (B); Rex Features: Alex Segre 7 (F); Clive Dixon 7 (A); Frederic Sierakowski 11 (motor cycle); Neale Haynes 42; Stewart Cook 68; Rock On Distribution: 15tl, 112l; Science Photo Library Ltd: 7 (D), 44 (Spark plug), 44 (apple); STILL Pictures The Whole Earth Photo Library: P Cairns 99; Vans Inc Ltd: 15tc, 15tr, 112c, 112r

All other images © Pearson Education

Cover images: *Front* iStockphoto: Kristian Stensoenes

Picture Research by: Kevin Brown

Every effort has been made to trace the copyright holders and we apologise in advance for any unintentional omissions. We would be pleased to insert the appropriate acknowledgement in any subsequent edition of this publication.

Designed by Keith Shaw

Cover design by Designers Collective

Project Managed by David Riley